ENVIRONMENTAL ACTION
Analyze Consider Options Take Action In Our Neighborhoods

ENERGY
Conservation

ENVIRONMENTAL ACTION
Analyze Consider Options Take Action In Our Neighborhoods

ENERGY
Conservation

A Student Audit of Resource Use

STUDENT EDITION

E2: ENVIRONMENT & EDUCATION

DALE SEYMOUR PUBLICATIONS®
Menlo Park, California

Developed by E2: Environment & Education™, an activity of the Tides Center.

Managing Editor: Cathy Anderson
Senior Editors: Jeri Hayes and Jean Nattkemper
Production/Manufacturing Director: Janet Yearian
Senior Production Coordinator: Alan Noyes
Design Manager: Jeff Kelly
Text and Cover Design: Lynda Banks Design
Art: Rachel Gage, Andrea Reider
Composition: Andrea Reider
Clip Art Illustrations: Copyright ©Art Parts, Courtesy Art Parts, 714-834-9166

This book is published by Dale Seymour Publications®,
an imprint of Addison Wesley Longman, Inc.

Dale Seymour Publications
2725 Sand Hill Road
Menlo Park, CA 94025
Customer Service: 800 872-1100

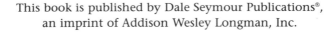

Printed on acid-free,
85% recycled paper
(15% post-consumer).
Printed using soy-based ink.

ISBN 0-201-49529-5
DS30674
1 2 3 4 5 6 7 8 9 10–ML–01 00 99 98 97

CONTENTS

Welcome to Environmental ACTION!

EXPLORE the Issues

ANALYZE

Act Locally

CONSIDER OPTIONS

Act Locally

TAKE ACTION

Appendices

Welcome to Environmental ACTION!

Welcome to Environmental ACTION!

This environmental program is designed to give you the knowledge and tools you need to make choices that will make a real difference to your quality of life, both now and in your future. You and all other living things modify the environment in order to live. What are the consequences of your actions? What is your impact on other living things and where they live? What is your impact on the food supply, atmosphere, and water cycle? The inter-relationships of living things and long-term effects of actions are only beginning to be understood. As human beings, we are unique among earth's organisms because we can choose to change our daily behavior. We can change our actions to reduce our impact on the environment, improve our quality of life, and provide for the needs of future generations. We can conserve and preserve our natural resources.

Using your school as a laboratory, you will investigate environmental issues and analyze how they influence human health and the environment. Each module contains a set of ACT activities that will guide you in your investigations. ACT stands for

- Analyze
- Consider Options
- Take Action

What features does a lifestyle with a sustainable future have?

It is renewable. Resources are replaced as they are used.
It is balanced. People and systems work together to improve the environment in the present and to ensure the quality of life in the future.
It is manageable. Products are reusable, recyclable, and biodegradable.

Your Journal

Throughout the project, you will be using a Journal. It is a notebook in which you record all your observations and data, write down ideas, make sketches, and outline procedures. You will need to use your Journal when you are conducting research in a study area, so it should be easy to carry. Your teacher may have specific instructions on what kind on notebook to use.

Action Groups

For most of the activities in this program, you spend part of your time working in a group. Your Action Group will work cooperatively, so that the group members benefit from each other's contributions. Sharing ideas, determining the best steps to take to achieve a goal, and dividing up tasks are just some of the advantages of working together.

Home activities can be done individually, but you may find that you prefer working with a group. Try to include your parents, brothers and sisters, or other family members in your work at home.

Topic Descriptions

The Environmental ACTION project that you are about to begin is one of six modules, or units. Each module focuses on a different aspect of the environment. Your teacher may choose to do only one module, a few modules, or all of the modules. The modules cover the following topics:

Energy Conservation

Using the school as a research laboratory, you'll explore where energy comes from and how it is used, the effect of energy production on the environment, and how to improve energy efficiency at school and at home.

Food Choices

You will investigate the effects of food production, diet, and nutrition on human health and the environment. You will analyze your school's food service programs and identify healthy choices and practices.

Habitat and Biodiversity

You will study the importance of biological diversity, landscape management, xeriscaping, composting, and integrated pest management (IPM). You'll tour the school grounds to assess the current landscaping lay-out and then evaluate the present condition in relation to environmental sustainability. This module also contains a step-by-step guide on how to create an organic garden and a seed bank.

Chemicals: Choosing Wisely

You will investigate the use of hazardous materials—paints, chemical products, cleaning supplies, pesticides—how they are stored and disposed of, and their potential effects on human health and the environment. After evaluating the results, you develop a plan for implementing the use of earth- and human-friendly alternatives at school and home.

Waste Reduction

After you sort your school's garbage to identify recyclable and compostable materials and analyze the school's current waste practices, you will formulate a plan to reduce your consumption and waste at school and at home. Development or improvement of a recycling program may be part of the process.

Water Conservation

After an introduction to water consumption and water-quality issues, you'll conduct an audit of water usage and efficiency to determine whether current consumption practices on campus can be improved. You will then develop strategies for implementing water conservation at school and home.

Explore the Issues

EVALUATE WORLD ENERGY USE

Energy is an essential natural resource. We could not exist without the heat, light, and food it provides. Find out about our increasing use of energy and how energy use differs throughout the world.

Setting the Stage

Discuss these questions:

1. What do you use energy for?

2. Why might people use different amounts of energy?

Vocabulary

→ fossil fuel
Btu

Focus

A. Study the graph and then discuss the questions. Material about fossil fuels can be found in Issues and Information section A.

WORLD POPULATION AND ENERGY USE

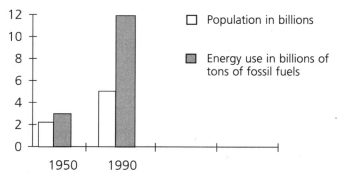

☐ Population in billions

▨ Energy use in billions of tons of fossil fuels

1. What was the change in population from 1950 to 1990? Did it double? Did it triple? Did it quadruple?

THINK ABOUT IT

"Since 1850 the human population has multiplied fivefold, from slightly over a billion people to five and a half billion. The use of energy per person... has increased fourfold. As a result, the total impact of humanity [on the use of the world's energy resources] has increased some twentyfold."

Paul R. & Anne H. Ehrlich quoted in Valerie Harms, et al., National Audobon Society, *Almanac of the Environment*, Grosset/Putnam, 1994, p. xiii

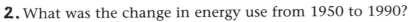

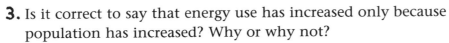
2. What was the change in energy use from 1950 to 1990?

3. Is it correct to say that energy use has increased only because population has increased? Why or why not?

B. Complete all the work on Activity Sheet 1.

It's a Wrap

On a separate sheet of paper, write a sentence to answer each question:

1. Why has world energy use increased so dramatically since 1950?

2. What might cause energy use to decrease worldwide?

Home

In your Journal, make a chart with these headings:

Time of Day	My Activity That Uses Energy	Energy Source

Fill in the first two columns by writing down each of your activities that depends on energy. Keep this record from the moment you leave the classroom until you get back to it the next day (or until 24 hours have passed). You will fill in the third column later.

Your teacher will give you an activity sheet like
the one below to use with this lesson.

EXPLORE

Name _____

WORLD ENERGY USE GRAPH

Make a bar graph showing all the information on the chart below.

1993 Population and Energy Consumption

Country	% of World Population	% of World Energy Consumed
U.S.	4.7	24.3
Brazil	2.7	1.8
U.K.	1.1	2.7
India	16.0	2.6
Japan	2.3	5.5
Russia	2.7	9.0

■ % Population
■ % Energy

25%
20%
15%
10%
5%
0%

U.S. Brazil U.K. India Japan Russia

1. Which country uses the most energy? _____

2. Which country has the most people? _____

3. Which countries have the highest energy use per person?

_____

4. Which countries have the lowest energy use per person?

_____

5. Why do you think energy use per person differs greatly from country to country?

_____

_____

IDENTIFY THE EFFECTS OF FOSSIL FUELS

EXPLORE

Fossil fuels—such as crude oil, coal, and natural gas—are the buried deposits of decayed plants and animals. These deposits have been affected, over hundreds of millions of years, by heat and pressure in the earth's crust. Burning fossil fuels provides most of the energy used in the U.S., but supplies are limited. They cannot be renewed and will not be replaced for thousands of years. Discover how the use of fossil fuels affects our environment and our health.

Setting the Stage

Discuss these questions:

1. What problems do fossil fuels cause to the environment and to our health?

2. How are fossil fuels used in your everyday life?

Vocabulary

acid rain
biomass
coal
crude oil
deforestation
landfill
natural gas
pollutant
smog

Focus

A. Review the material in Issues and Information section A. Then study the chart and discuss the questions on page 14.

THINK ABOUT IT

"If we continue consuming oil at present rates, known reserves will be depleted in about 40 years. Allowing for as yet undiscovered oil resources, the time span is increased by about another 30 years."

from Dr. Norman Myers, ed., *Gaia: An Atlas of Planet Management,* Anchor Books, Doubleday, 1993, p. 107

Main Air Pollutants, Human-Caused Sources, and Environmental Effects

Pollutant*	Source	Global Warming	Acid Rain	Smog	Damage to Vegetation, Wildlife, Humans
CO_2	Fossil fuels, deforestation	X			X
CH_4	Rice fields, cattle, landfills, fossil fuels	X			
NO, NO_2	Fossil fuels, biomass burning		X	X	X
N_2O	Nitrogenous fertilizers, deforestation, biomass burning	X			
SO_2	Fossil fuels, ore smelting		X		X
CFCs	Aerosol sprays, refrigerants, solvents, foams	X			X
O_3	Fossil fuels	X		X	X

*CO_2 = Carbon Dioxide, CH_4 = Methane, NO = Nitric Oxide, NO_2 = Nitrogen Oxide, N_2O = Nitrous Oxide, SO_2 = Sulfur Dioxide, CFCs = Chlorofluorocarbons, O_3 = Ozone

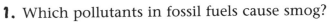

1. Which pollutants in fossil fuels cause smog?

2. What are some of the causes of global warming?

3. Which air pollutants listed on the chart are not present in fossil fuels?

4. Which of the environmental effects listed on the chart are causing human health problems today? What are they?

B. Complete all the work on Activity Sheet 2. Look at Issues and Information section B for information if you need help answering the questions.

It's a Wrap

On a separate sheet of paper, write a sentence to answer each question:

1. How are fossil fuels damaging to the environment?

2. How are fossil fuels damaging to our health?

3. If more nonpolluting energy sources are used, will our environment get better?

4. How can we reduce global warming?

Home

 In your Journal, look at the chart you began in Activity 1. Write "fossil fuel" in column 3 for each activity that uses oil, natural gas, or coal as an energy source. Then identify the type of fossil fuel used.

Your teacher will give you an activity sheet like
the one below to use with this lesson.

Name _____

U.S. ENERGY USE

Use the information in the table below to make a pie chart, and then answer the questions.

Energy Use in the United States

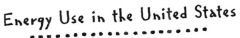

Source	Percentage
Oil	42
Natural gas	25
Coal	23
Hydroelectric	5
Nuclear	4
Other	1

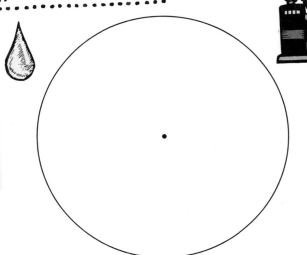

1. What is the percentage of energy provided by fossil fuels? By other sources?

_____

_____

2. When the fossil fuels are used up, what sources will supply our energy needs?

_____

3. What sources of energy might be included under the heading "Other"? Look at Issues and
Information section B for information if you need help.

_____

_____

_____

COMPARE RENEWABLE AND NONRENEWABLE ENERGY RESOURCES

Renewable energy resources are those that can be replaced through natural processes. Nonrenewable energy resources, such as fossil fuels, are those that cannot be replaced. Learn how to tell the difference. Discover ways to use renewable sources at home and at school.

Setting the Stage

Discuss these questions:

1. What are some renewable energy resources?
2. How can you encourage more use of renewable energy resources and less use of nonrenewable ones?

Vocabulary

biogas
geothermal
hydroelectric
photovoltaic
solar thermal
turbine

Focus

A. Read the materials about Renewable Energy Resources in Issues and Information section B. Then study the chart and discuss the questions.

1. Which renewable energy resources were not in general use before 1984?
2. Which renewable energy resources were being used at least somewhat by 1984?
3. Why did this use of renewable energy resources change?
4. According to the chart, which two renewable energy resources are expected to be used most in the future?

THINK ABOUT IT

"Historically, wind and water have been important sources of energy. In the U.S., over 8 million mechanical windmills have been installed since the 1860s: in the 1920s and 1930s …many of these were wind generators, each delivering 200 to 300 watts, and providing all the electrical needs of outlying farms at the time."

from Edward Harland, *Eco-Renovation*, Chelsea Green Publishing Co., 1994, p. 47

Production of Selected Renewable Energy Sources in the United States, 1984-2000

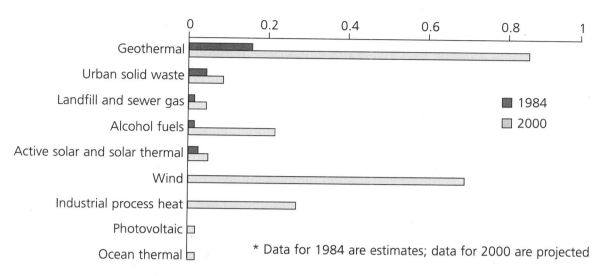

* Data for 1984 are estimates; data for 2000 are projected

B. Complete all the work on Activity Sheet 3.

It's a Wrap

On a separate sheet of paper, write a sentence to answer each question:

1. Why do you think wind energy is not more widely used?

2. What might be some negative results of using biomass energy?

3. What is the one energy source that is available all over the world?

Home

In your Journal, look at the chart you began in Activity 1 and continued in Activity 2. Make a separate list of all your activities that use fossil fuels. Next to each item, write a renewable energy resource that you might be able to use instead.

Your teacher will give you a two-part activity sheet
like the one below to use with this lesson.

ACTIVITY SHEET

Name

ENERGY RESOURCES (part 1)

Study the drawing below. Draw a circle around each renewable energy resource that you find.
Draw a box around each nonrenewable energy resource that you find. Then answer the questions.

PRACTICE CONSERVATION

EXPLORE 4

Conservation is the responsible use of natural resources. It helps us pass along to future generations most of the same resources we use today. Discover how important it is to practice energy conservation as a way to solve present and future energy problems.

Setting the Stage

Discuss these questions:

1. How can you practice energy conservation in your every-day life?

2. What are the benefits of practicing energy conservation?

Vocabulary

conservation
daylighting
fluorescent
kilowatt-hour (kWh)
lumen
incandescent
watt

Focus

A. Study the pictures on p. 20. Then discuss the questions.

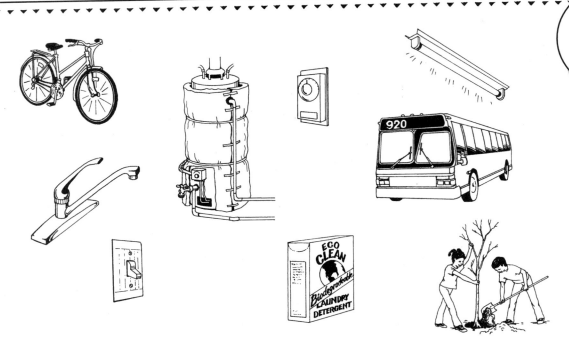

1. How can you conserve energy when you are at home?
2. How can you conserve energy when you are at school?
3. Name some ways to conserve energy that are not shown in the pictures.

B. Complete all the work on Activity Sheet 4. You may wish to look at Issues and Information section C.

It's a Wrap

On a separate sheet of paper, write a sentence to answer each question:

1. What are two ways that an energy-efficient appliance helps conserve our natural resources?
2. What are some ways to encourage people to use energy-efficient appliances in the future?

Home

In your Journal, look at the chart you have been working on. Make a fourth column labeled "How I Could Conserve Energy." Write a conservation suggestion for each activity.

Your teacher will give you an activity sheet like
the one below to use with this lesson.

Name

BULB COMPARISON

Read the information about light bulbs. Then complete the chart below.

	incandescent bulb	vs	compact fluorescent bulb
Average life of bulb	1000 hours		10,000 hours

Thirteen 100-watt incandescent bulbs equal one 27-watt fluorescent bulb

- Fluorescent bulbs are four times more efficient than incandescent bulbs, and they last up to 13 times longer.

- A 100-watt incandescent bulb burned for an average of 5 hours a day uses 0.5 kilowatt-hours of electricity per day.

- An energy-efficient 27-watt compact fluorescent bulb gives off the same light as a 100-watt incandescent bulb. If it is burned for an average of 5 hours a day, it uses 0.135 kilowatt-hours of electricity per day.

- For each kilowatt-hour of electricity used, one-and-a-half pounds of carbon dioxide (CO_2) enter the atmosphere.

- The average cost for electricity in the U.S. is 8.76 cents per kilowatt-hour.

	One 100-watt incandescent bulb	One 27-watt fluorescent bulb	Savings for one bulb	Savings for 100 bulbs
kWh used per year (average 5 hrs./day)				
Pounds of CO_2 put into the atmosphere per yr.				
Energy cost per year (kWh x .0876)				

Analyze

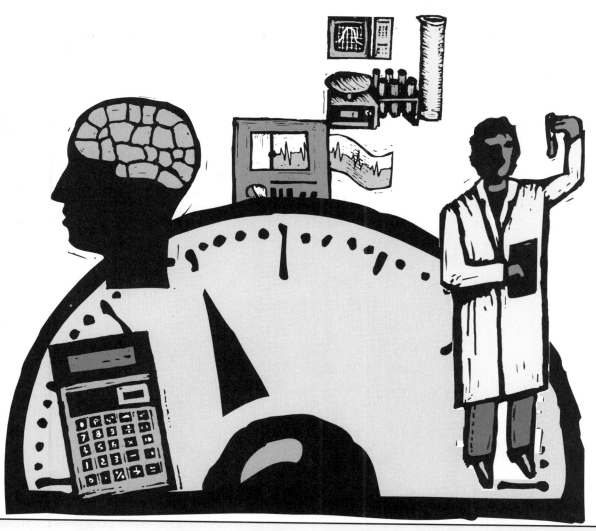

TOUR SCHOOL ENERGY SITES

If you want to encourage wise use of energy in your own environment, a good place to start is your school. Take a tour of the school. A school staff member or community resource person will help you find out how energy is being used.

Setting the Stage

Discuss these questions:

1. What is energy used for in school?

2. Where does this energy come from?

Vocabulary

 thermostat

Focus

To find out about your school, use Activity Sheet 5 as a guide for asking questions. The answers you get will help you fill out the chart. Make notes on the chart or on a separate sheet of paper.

Locate energy-using systems—such as heating and cooling—and equipment—such as computers, printers, copiers; refrigerators and other appliances; indoor and outdoor lighting. Find out about how the equipment and systems are used and what energy sources make them work. For example, what is used to provide heat? Is it natural gas? Is it oil? Is it coal? School staff can be a resource to help answer questions. You may also refer to Issues & Information section F.

> ### THINK ABOUT IT
>
> "If you are thinking a year ahead, sow seed. If you are thinking ten years ahead, plant a tree. If you are thinking 100 years ahead, make people aware. By sowing seed once, you will harvest once. By planting a tree, you will harvest tenfold. By opening the minds of people, you will harvest one hundredfold."
>
> Chinese proverb

It's a Wrap

 Discuss your work on Activity Sheet 5 with your classmates. Add or change items as you receive feedback and suggestions. In your Journal, write a short paragraph telling how the tour helped you find out how energy is used in school and where it comes from.

Home

 In your Journal, set up charts with the same headings as the ones on Activity Sheet 5. Fill out the charts to show how energy is used in your home.

Your teacher will give you a two-part activity sheet like the one below to use with this lesson.

ACTIVITY SHEET

ANALYZE

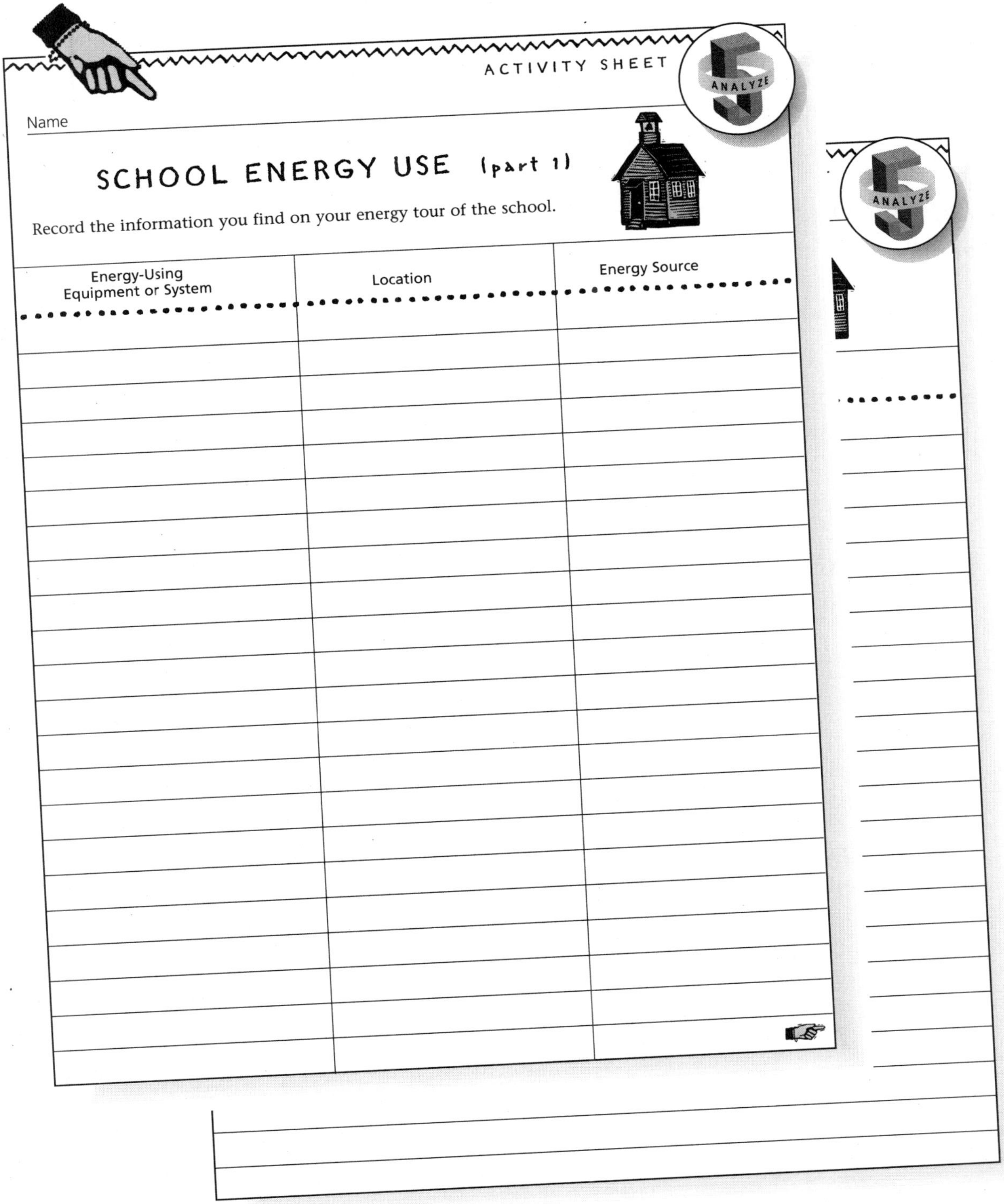

Name

SCHOOL ENERGY USE (part 1)

Record the information you find on your energy tour of the school.

Energy-Using Equipment or System	Location	Energy Source

MEASURE ENERGY USE

To complete an Energy Audit of your school, you need to know how to measure energy use. You also need to know how much the school pays for energy. Practice reading meters and energy bills.

Setting the Stage

Discuss these questions:

1. How is energy use measured?

2. How do energy providers know what to charge the school?

3. How can you compare energy use from one day to the next, one season to the next, or one year to the next?

Vocabulary

 therms

Focus

A. Gas meters usually have four main dials. Each dial is numbered from 0 to 9. (The small dials with no numbers are used for test purposes only.) The numbers on the dials show units of gas measured in hundreds of cubic feet. Start with the left dial and read across to the right. When the pointer falls between two numbers, use the lower number. Note that 0 can either be higher than 9 or lower than 1. Look at the gas meters on the next page.

THINK ABOUT IT

"Approximately 8% of urban electrical demand in cities is used to cool buildings, cities being hotter than adjacent rural areas, due to large areas of concrete and asphalt. ...Deciduous vines and trees... provide summer shade, yet allow winter sun. ...An average front lawn has the cooling effect of 10 tons of air conditioning."

from Lawrence Livermore Laboratory, UC Berkeley; Sacramento Municipal Utility District; and Dr. James B. Beard, quoted in The Council for a Green Environment's *Greening the Urban Ecosystem*, p. 2

What is the measurement shown on the dials? Write your answer on a separate sheet of paper.

B. Complete Activity Sheet 6 to practice reading an electric meter.

C. Issues and Information section C will help you understand how energy is measured if you need help. In section D you will find information on how to understand and interpret the information included on standard utility bills. To answer the questions below, you will need to look at a utility bill from your school or from home. Your teacher will help you locate a sample school utility bill, or ask volunteers to bring some from home. On a separate sheet of paper, answer the following questions:

1. What is the energy use in therms or kilowatt-hours for the billing period?

2. What is the rate per therm or per kilowatt-hour?

3. How much money would be saved if there were a 10% reduction in energy use for the billing period?

4. What would the total energy savings be if there were a 10% reduction in energy use for the billing period?

5. Why might energy costs change from one season to another or from one year to another?

It's a Wrap

Review Activity Sheet 6 with your classmates. Then put your knowledge to work:

You work for a utility company. One of your business customers has a plan for cutting down on energy use but needs some questions answered first. How can your customer tell whether or not the plan is working? What should the customer do before putting the plan into action? Write your answers on a separate sheet of paper.

Home
· · · · ·

Check the utility meters at your home every day for a week. Read the meters at the same time every day. Record the date and reading in your Journal. If possible, check your family's utility bills over the past year to determine seasonal differences in energy use. (Some companies may include a summary and comparison of the year's energy use on every bill.) How much money would your family save if home energy use were reduced by 10%?

Your teacher will give you an activity sheet like
the one below to use with this lesson.

Name _____

READING A METER
AND CALCULATING COST

The numbers on the dials of an electric meter are units of electric energy called
kilowatt-hours, or kWh. Each dial moves in an opposite direction from the one next
to it and is read like a one-handed clock. The dials are read from left to right. If a
pointer falls between two numbers, use the lower number. Note that 0 can either be
higher than 9 or lower than 1.

Write the numbers for each reading on the lines.

First Reading   Kilowatt-Hours

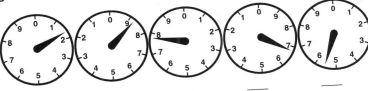

____ ____ ____ ____ ____

Second Reading

____ ____ ____ ____ ____

CALCULATE THE COST

First find the number of kilowatt-hours by subtracting the first meter reading from
the second. Then multiply the number of kilowatt-hours by the cost per kWh.

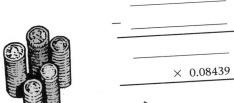

_____	Second meter reading
− _____	First meter reading
_____	kWh electricity used
× 0.08439	cost/kWh (summer rate)
$_____._____	Cost

CUT ENERGY USE

Now that you know how to measure energy use, it is important to know how to cut energy use. There are many different ways to save, or conserve, energy.

Setting the Stage

Discuss these questions:

1. How is energy wasted?
2. How can energy be conserved?

Vocabulary

caulking
glaze (single-glazed, double-glazed)
insulate
weather-stripping

Focus

A. One way to find out how to conserve energy is to look at equipment or systems where energy is lost. Think of a few examples from home or school and write them on a sheet of paper. Then write down three actions that each person could take to save energy at home or at school. Remember, energy is wasted in many different ways, including our own habits. You may also look at the ideas presented in Issues and Information sections E, F, and G.

B. Complete Activity Sheet 7. Share your ideas with your classmates. During the class discussion, you may change or add to what you put on the sheet.

THINK ABOUT IT

"Caulking and weather-stripping are the easiest and least expensive weatherization measures and can save more than 10% on energy bills.... [They] are most often applied to doors and windows, which account for about 33% of a home's total heat loss. Because windows out-number doors, energy efficiency features of windows are particularly important to lowering energy costs."

from "A Guide to Making Energy-Smart Purchases," *Energy Efficiency and Renewable Energy Clearinghouse,* April 1994, p. 2

It's a Wrap

Draw a cartoon or simple sketch to show energy loss. Add a caption to explain how energy might be conserved rather than wasted in the situation you illustrated.

Home

Look around your home. Find places where energy is lost. Make a list in your Journal. Pay attention to family members' actions. Take notes on how their actions affect energy use in your home.

Your teacher will give you a two-part activity sheet like the one below to use with this lesson.

ACTIVITY SHEET

Name

WAYS TO CONSERVE ENERGY (part 1)

1. Draw a circle around each area where energy might be wasted or around items that might waste energy.

erved

PLAN YOUR ENERGY AUDIT

ANALYZE

You and your classmates will form Action Groups to conduct a school Energy Audit. Each Action Group will investigate an area of the school or an energy system. This activity will help you focus your Energy Audit and assign tasks.

Setting the Stage

Discuss these questions:

1. According to your findings recorded on Activity Sheet 5, what equipment uses energy at the school?
2. According to your findings recorded on Activity Sheet 5, what systems use a lot of energy at the school?
3. What area or system would benefit most from an energy audit that explores ways to conserve energy?

Focus

A. Think of all the different energy-using areas of your school—classrooms, cafeteria, auditorium, gymnasium, media center, office, and so on. In your Journal, make a heading for each location. Under each heading, write down the energy system or equipment used in that location. Then find a way to categorize the energy systems and equipment.

B. Discuss how the Action Groups will organize the Energy Audit. Will you investigate a particular system or type of equipment schoolwide, a particular system or type of equipment in a certain area, or all the systems and equipment in a certain area?

C. Once you have decided what you will explore, discuss how you will break the investigation into smaller tasks that can be assigned to each Action Group.

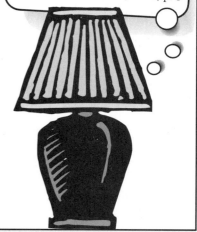

THINK ABOUT IT

"Lighting accounts for 20–25 percent of all electricity sold in the United States.... If energy-efficient lighting were used..., the nation's demand for electricity would be cut by more than 10 percent. This would result in annual reductions of 202 million metric tons of carbon dioxide—the equivalent of taking 44 million cars off the road; 1.3 million metric tons of sulfur dioxide; and 600,000 metric tons of nitrogen oxides. "

from "Green Lights: An Enlightened Approach to Energy Efficiency and Pollution Prevention," U.S. Environmental Protection Agency, July 1993, p. 3

D. Meet briefly with your Action Group to brainstorm how you will complete your assigned task. Think about what information you need for your Energy Audit. List specific questions you need to answer as you conduct your Energy Audit in Activity 9. Choose one member of your Action Group to take notes and then work together to fill out Activity Sheet 8.

It's a Wrap

In your Journal, record reasons behind the choice your class made for its Energy Audit. What were the advantages and disadvantages of the other choices you discussed in class?

Home

Think of a particular home energy system or appliance you can investigate for an Energy Audit. In your Journal, write down your ideas and decide which one would benefit most from an investigation. Determine what information you will need and how you will go about finding it. Get ideas from family members and obtain your parents' permission to carry out an Energy Audit once you have a plan.

Your teacher will give you an activity sheet like
the one below to use with this lesson.

ACTIVITY SHEET

ANALYZE

Name _____

Action Group _____

ENERGY AUDIT PLAN

Name	Task	Resources Needed (People/ Equipment)	Permission Required/ Obtained	Report Due Date

CONDUCT YOUR ENERGY AUDIT

Your Action Group can work together to conduct the Energy Audit. Developing a list of questions can serve as a guideline for finding information so that each group member can complete his or her part of the task.

Setting the Stage

Discuss these questions:

1. What area or system or type of equipment will you investigate?

2. What information do you need to find out to get started?

Focus

A. Work with your Action Group to brainstorm questions that will help you find out what you need to know for your Energy Audit. For example, for your area, system, or equipment:

1. What energy source is used?

2. When is it in use?

3. How do on/off switches or master controls work? Are they centrally located, or are there separate controls throughout the school?

4. What other energy source could be used either in addition to or instead of what is used?

5. How is it cleaned and maintained?

6. What device—such as insulation or a timer—could be installed to cut energy use?

7. What could staff and students do to cut energy use?

THINK ABOUT IT

"Insulating is the most important of all energy-conserving measures, because it is with the use of insulation that we can have the greatest impact on our energy expenditure: for the average house we can reduce the amount of heat lost through the fabric of the house by at least half...."

from Edward Harland, *Eco-Renovation*, Chelsea Green Publishing, p. 57

B. Look at Activity Sheet 8 and your list of questions to help you figure out where to start your audit and what steps you need to take. If necessary, add names of people you need to talk to or tasks that need to be done that were not included on Activity Sheet 8.

C. Use the forms on Activity Sheet 9 to keep track of your findings as you conduct an Energy Audit of the area, system, or equipment you chose.

It's a Wrap

In your Journal, write a summary of your Energy Audit findings and recommendations. What information surprised you the most? What were your observations or overall impressions of how energy is used at school?

Home

In your Journal, set up an information sheet similar to Activity Sheet 9 to use as you conduct your Energy Audit at home. Interview family members to find out more information about how equipment is used and maintained in your home, and take notes about energy waste and energy savings that you observe.

Your teacher will give you a two-part activity sheet
like the one below to use with this lesson.

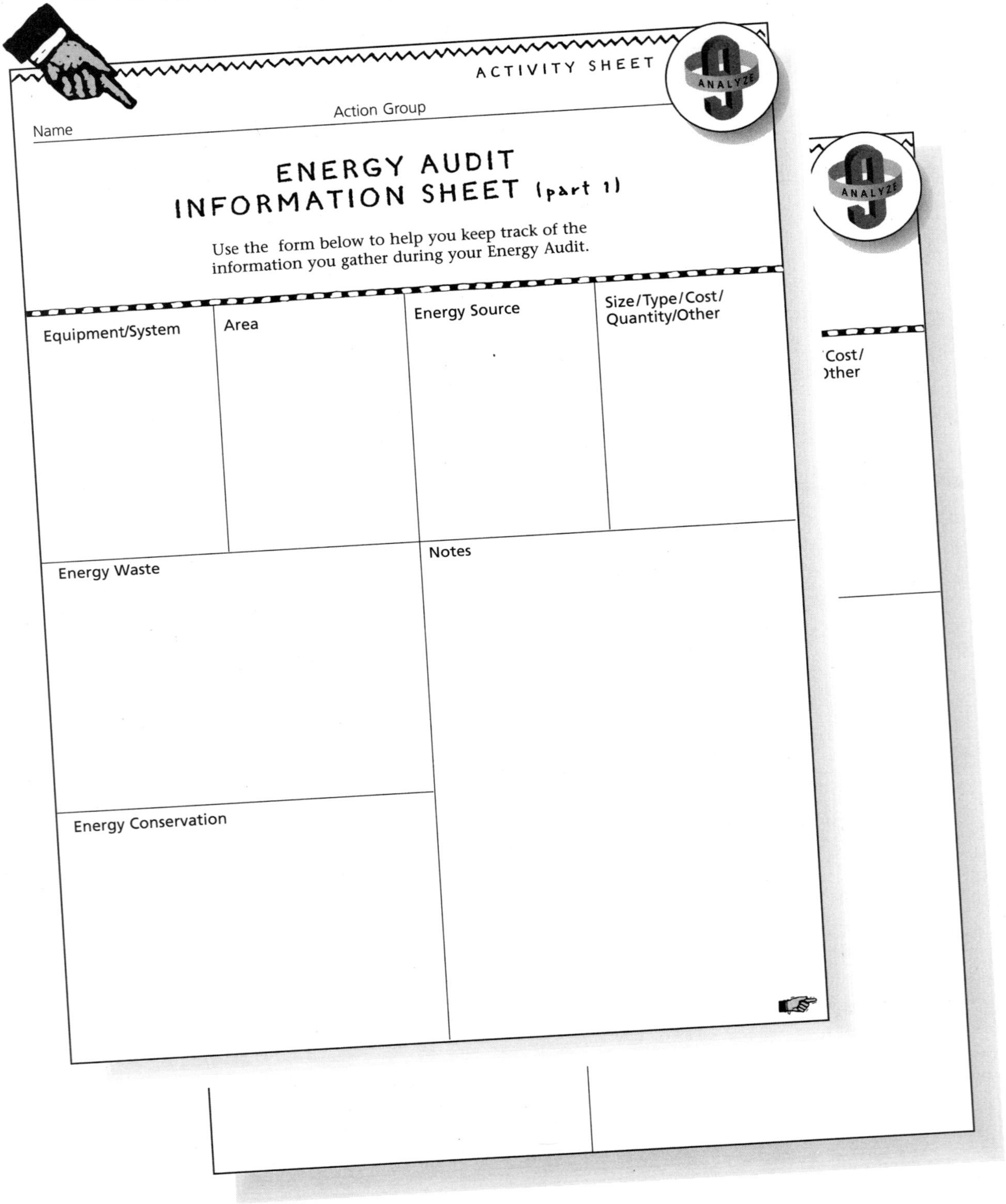

ACTIVITY SHEET

Action Group

Name

ENERGY AUDIT
INFORMATION SHEET (part 1)

Use the form below to help you keep track of the
information you gather during your Energy Audit.

Equipment/System	Area	Energy Source	Size/Type/Cost/Quantity/Other

Energy Waste	Notes

Energy Conservation	

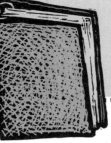

SUMMARIZE FINDINGS

Y ou are ready to put together all of the information collected by your Action Group. Your group will then use the information to summarize the results of your Energy Audit and share them with other Action Groups.

Setting the Stage

Discuss these questions:

1. What have you learned about energy use in the school?

2. How can you present the information so that it can easily be understood and interpreted?

Focus

A. Meet with your Action Group. Use the information you gathered on Activity Sheet 9 to help you fill in the chart on Activity Sheet 10. Activity Sheet 10 will combine the findings, conclusions, and observations of everyone in your group.

B. Get together with your Action Group. Using Activity Sheet 10 as a guide, write a one-page report to describe the results of your Energy Audit. Present your final report to the class.

C. Compare the findings that your group members collected to the summaries that other Action Groups have prepared. Discuss any differences. You may wish to make changes in your summary after the discussion.

THINK ABOUT IT

"...converting all of the nation's exit signs to energy-efficient systems would save an estimated 2.7 billion kWhs of energy annually— the equivalent of planting 780,000 trees, taking 380,000 cars off the road, or saving 260,000,000 gallons of gasoline."

from "Green Lights Update," U.S. Environmental Protection Agency, October 1994, p. 6

It's a Wrap

Discuss the Energy Audit, taking into account the summaries of all the Action Groups. Use your Journal to record the answers to the following questions. Explain your findings for each answer.

1. Which systems or equipment in the school uses the most energy? Which uses the least energy?
2. Which system or equipment is the most energy efficient? Which wastes the most energy?
3. What else did the class discover about energy use during its Energy Audit?

Home

After you have completed your home Energy Audit, answer the following questions. Write your answers in your Journal.

1. Which system or equipment at home uses the most energy? Why?
2. Which system or equipment at home uses the least energy? Why?
3. Which system or equipment at home wastes the most energy? How?
4. Which system or equipment is most energy efficient? Why?

Your teacher will give you an activity sheet like the one below to use with this lesson.

ACTIVITY SHEET

Name _____

Action Group _____

ENERGY AUDIT SUMMARY

Equipment/System	Area	Energy Source	Size/Type/Cost/ Quantity/Other	Energy Waste	Energy Conservation

ACT LOCALLY

Now that you have explored a wide variety of energy-saving ideas, share the best of the bunch with your schoolmates by carrying out a project such as the following:

A. Create an article for the school paper or a school-wide flyer featuring a list of conservation hints that students can follow at school and at home.

B. Join in a community effort such as a conservation walk-a-thon, park cleanup, or tree planting.

C. Help plan a Conservation Conversation evening. Invite parents and friends to attend and find out about your energy-saving activities.

D. Present your conservation information to your school Parent-Teacher group and explain the Energy Audit your Action Groups are undertaking. The organization may be able to help you implement some of your ideas.

THINK ABOUT IT

"If every U.S. household lowered its average heating temperature by 6°F for a day, we'd save 500,000 barrels of oil. If we were to put on another sweater instead of cranking up the heater, the benefits would be cleaner air and a healthier future for all of us."

from Robbins & Solomon, *Choices for Our Future*, Book Publishing Company, 1994, p. 109

Consider Options

BRAINSTORM CONSERVATION IDEAS

CONSIDER OPTIONS

Conserving energy can depend on coming up with a bright idea or changing an old habit. Use what you have learned in your research to brainstorm conservation ideas and share them with your classmates.

Setting the Stage

Discuss these questions:

1. How can energy awareness lead to effective conservation strategies?

2. What are some factors to consider if you are finding ways to conserve energy?

Focus

You have been exploring how different sources of energy are used, how they are wasted, and how they can be conserved. For example, fossil fuels are used to heat a room. How is this heat wasted? How can it be conserved? What can you do to help conserve heat? Use the chart on Activity Sheet 11 to record your ideas. You may wish to look at Issues and Information section E if you are stuck.

It's a Wrap

In a class discussion, share your ideas about conserving energy. If you pick up some bright ideas from your classmates or think of new ideas during the discussion, add them to your chart. Go back and star the ideas that depend on changing behavior or habits.

Home

Make a chart to record energy-saving ideas for wasted energy in your home. Be sure to include steps family members can take to make appliances run more efficiently, as well as ways they can cut energy use.

THINK ABOUT IT

"...almost every poll shows Americans decisively rejecting higher taxes on fossil fuels, even though that proposal is one of the logical first steps in changing our policies in a manner consistent with a more responsible approach to the environment."

Senator Al Gore, *Earth in the Balance*, Houghton Mifflin Co., 1992, p. 173

Your teacher will give you an activity sheet like
the one below to use with this lesson.

Name _____   Action Group _____

ENERGY CONSERVERVATION IDEAS

Use the chart below as you brainstorm ideas about conserving energy around the school.
Record your ideas in the appropriate boxes. Be sure to include ideas for maintenance and
repairs, as well as ways to cut down on energy use.

 Electricity

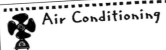 Air Conditioning

 Light

 Heat

 Appliances

WEIGH CONSERVATION COSTS AND BENEFITS

CONSIDER OPTIONS
12

Solutions to energy problems may have surprise fringe benefits or unexpected hidden costs. Evaluate costs and benefits of conservation measures thoroughly and then share your findings.

Setting the Stage

Discuss these questions:

1. What factors need to be considered when you are deciding which conservation measures to implement?

2. Why is it important to consider long-range benefits and look for hidden costs when deciding which conservation measures to implement?

Focus

You have been discovering ways to conserve energy here at school. When you consider a conservation measure, you need to figure out what the costs and benefits will be to put your idea in motion. Ask yourself the following questions:

1. What are the monetary costs?

2. What are the nonmonetary costs?

3. Are there any hidden costs? What are they?

4. What are the immediate benefits?

5. What are the long-term benefits?

6. Are there unexpected fringe benefits to the environment, to quality of life, to health, and so on?

Pose these kinds of questions as you work on your own or with a partner to evaluate your Action Group's energy-saving ideas. Record your findings on Activity Sheet 12.

THINK ABOUT IT

"We live in a world where energy has never been so cheap and easy to use. This has led us to waste it on a massive scale.... [The] real ecological costs of energy ...[—]global warming, resource depletion, and acid rain[—]are impossible to calculate, as we simply do not know what their final effects will be."

from Edward Harland, *Eco-Renovation*, Chelsea Green Publishing, p. 37

It's a Wrap

Share with your Action Group what you discovered about the
hidden costs and unexpected benefits of energy-saving measures.
If additional costs or off-setting benefits are mentioned, factor them
into your evaluation or highlight them for further consideration.

Home

Copy the form from Activity Sheet 12 into your Journal. Use it
to outline the costs and benefits of energy-saving ideas you and
your family are considering. Then evaluate your ideas.

Your teacher will give you a two-part activity sheet
like the one below to use with this lesson.

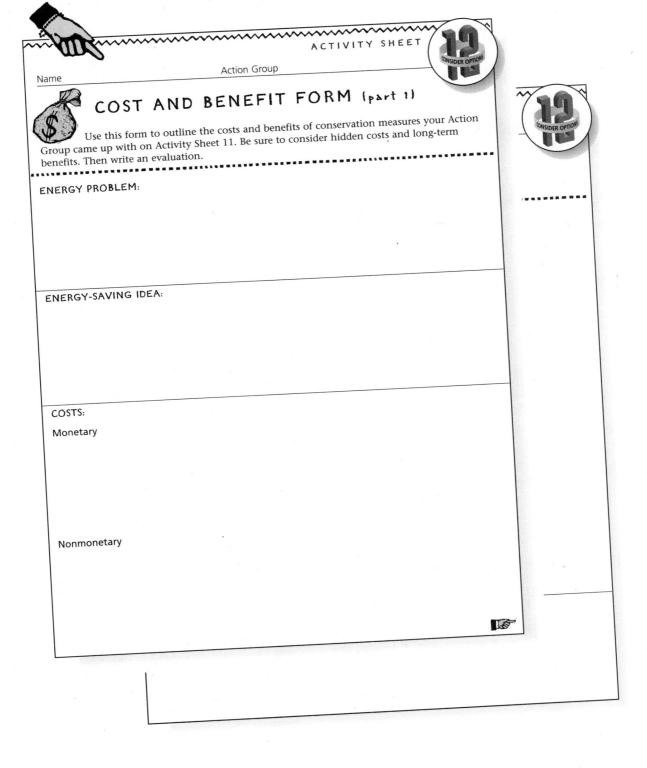

ACTIVITY SHEET

Name _____ Action Group _____

COST AND BENEFIT FORM (part 1)

Use this form to outline the costs and benefits of conservation measures your Action
Group came up with on Activity Sheet 11. Be sure to consider hidden costs and long-term
benefits. Then write an evaluation.

ENERGY PROBLEM:

ENERGY-SAVING IDEA:

COSTS:

Monetary

Nonmonetary

MAKE CONSERVATION RECOMMENDATIONS

A thorough investigation and assessment of conservation measures has given you a clearer view of the big picture. Using what you have learned, you can recommend which measures to adopt.

Setting the Stage

Discuss these questions:

1. What are the most important considerations to keep in mind as you explore conservation measures for the school?

2. Are low-cost conservation measures always the best choice? Explain your answer.

Focus

Your Action Group has been weighing the costs and benefits of conservation measures to implement here at school. Now you are going to make recommendations based on your analysis, choosing the best energy-saving plans and deciding how to present them to the class. What information about your ideas should you include? What arguments will be most persuasive? How can you present convincing details? Use Activity Sheet 13 to plan your presentation.

It's a Wrap

In a class discussion, outline some of the ideas your Action Group eliminated as you assessed your conservation plans and explain your reasoning. Classmates may have ideas for modifying or reorganizing these rejected ideas to make them more feasible, beneficial, or cost effective.

THINK ABOUT IT

"The energy efficiency of similar appliances can vary significantly.... Determining and comparing the energy efficiency of different models is usually easy, because federal regulations require many types of appliances to display EnergyGuide labels.... EnergyGuide labels indicate either the annual estimated cost of operating the system or a standardized energy efficiency ratio."

from "A Guide to Making Energy-Smart Purchases," *Energy Efficiency and Renewable Energy Clearinghouse,* April 1994, p. 4

Home

 Look at the conservation ideas you evaluated for yourself and your family. Take into account practical considerations and the cooperation of family members as you decide on which recommendation to present.

Your teacher will give you a two-part activity sheet
like the one below to use with this lesson.

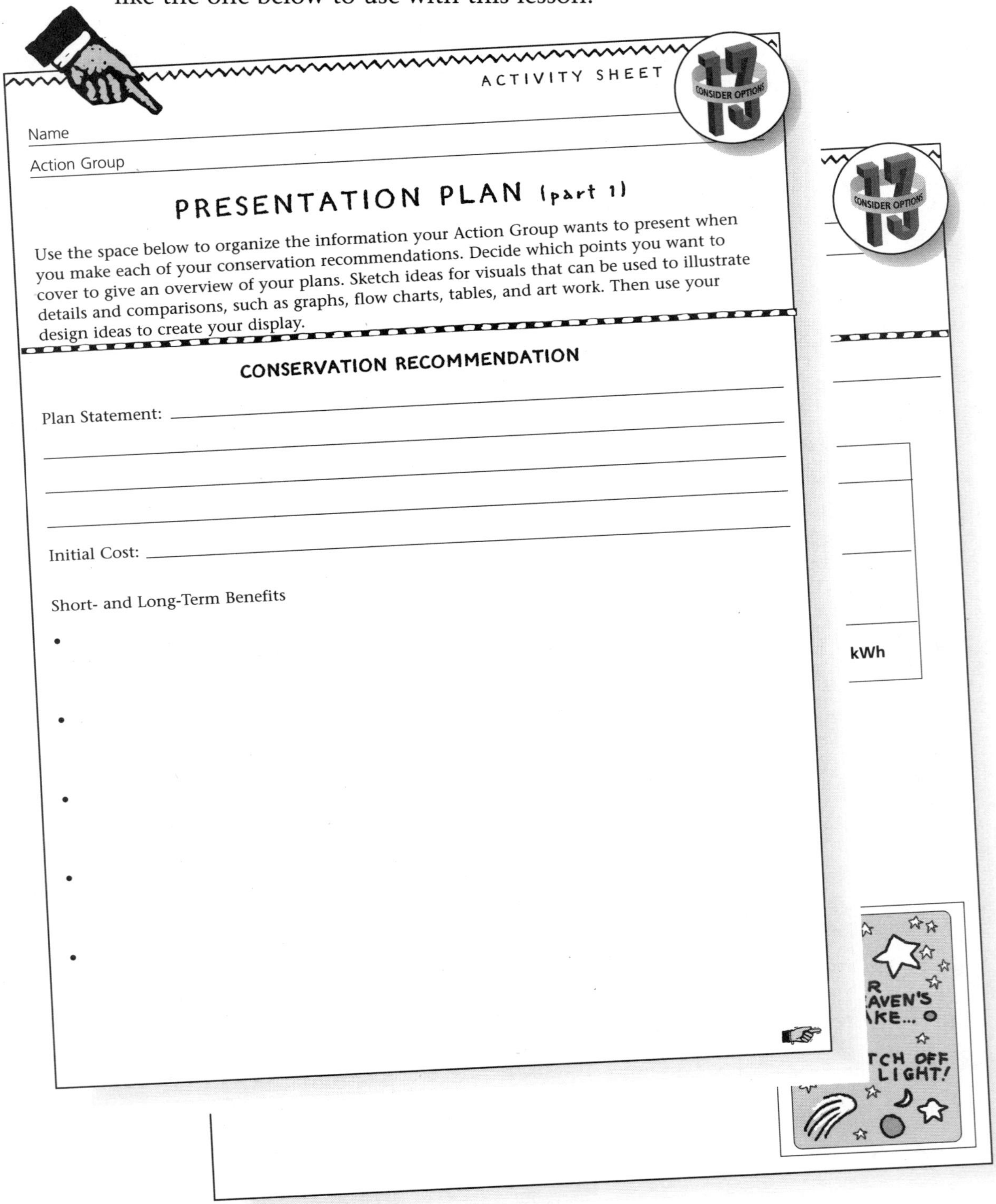

ACTIVITY SHEET

Name _____

Action Group _____

PRESENTATION PLAN (part 1)

Use the space below to organize the information your Action Group wants to present when
you make each of your conservation recommendations. Decide which points you want to
cover to give an overview of your plans. Sketch ideas for visuals that can be used to illustrate
details and comparisons, such as graphs, flow charts, tables, and art work. Then use your
design ideas to create your display.

CONSERVATION RECOMMENDATION

Plan Statement: _____

_____

_____

_____

Initial Cost: _____

Short- and Long-Term Benefits

•

•

•

•

•

kWh

ACT LOCALLY

ACT LOCALLY

Now that you have a better understanding of how energy is used at your school, find ways to share what you have learned. Think of a project that will help increase conservation awareness and involvement, or use one of the suggestions given below.

A. Invite parents and students to a demonstration of Everyday Ways to Save Energy. Include a display that shows energy-conserving equipment.

B. Hold a school contest or raffle with an energy-conservation theme. Award prizes such as T-shirts, books, posters, or bumper stickers with slogans that encourage conservation awareness.

C. Present your Energy Audit summary to a local business or community organization. Use your own findings to suggest ways in which businesses or agencies can work to conserve energy, including the purchase of energy-saving equipment, maintenance and repair of existing systems and equipment, strategies to increase awareness and promote changes in behavior, and the recycling and reuse of office materials.

THINK ABOUT IT

"The average American is responsible for the emission of 55,000 pounds of carbon dioxide and its equivalent in other greenhouse gases every year, mainly through the use of energy."

from Valerie Harms, et al., National Audubon Society, *Almanac of the Environment,* Grosset/Putnam, 1994, p. 59

Take Action

CHOOSE CONSERVATION MEASURES

14 TAKE ACTION

Your Action Group has decided on its conservation recommendations and prepared a presentation. Now your Action Group will make its pitch to classmates.

Setting the Stage

Discuss these questions:

1. What factors will help put an energy-saving idea into action?

2. What factors will help make an energy-saving measure successful?

3. How do energy-saving measures for your family compare to energy-saving measures for your school?

Focus

Each Action Group is going to make its presentation to the class, outlining its energy-saving plans. Use Activity Sheet 14 to make notes and rate each group's ideas on a scale of 1 to 3. Also jot down any questions you want to discuss further as you decide which ideas are the best. Once all the presentations have been made and evaluated, you and your classmates will reach a consensus on which measures to propose to the school committee.

It's a Wrap

Review the information presented, along with your impressions of and opinions about the conservation measures you will propose. Write a paragraph or draw a cartoon illustrating why your conservation measure will succeed.

THINK ABOUT IT

"How can we make the very best use of sunlight...? It is useful to remember that daylight, when entering through a standard-sized, unimpeded window during the middle of the day, can supply the lighting equivalent of between five and thirty 100-watt bulbs."

from Edward Harland, *Eco-Renovation*, Chelsea Green Publishing, p. 94

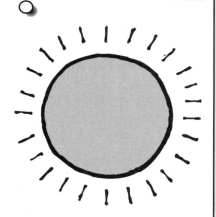

Home
•••••

Present energy-saving ideas to your family. Have family members rate each idea according to how much energy it will save, how much time it will take, how each family member can participate, how much setting up the idea will cost, and so on. Use the rankings to reach a consensus about which measures to adopt.

Your teacher will give you an activity sheet like the one below to use with this lesson.

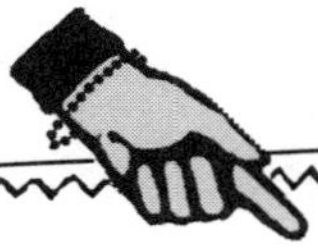

ACTIVITY SHEET

14 TAKE ACTION

Name

RATING SHEET

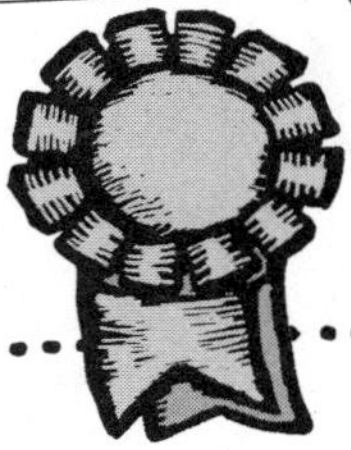

Fill in the following rating sheet for each presentation.

Group

Plan

Costs

Low • • • High

Benefits

Low • • • High

Energy Savings

Low • • • High

Cooperation Incentives

Low • • • High

Long-Term Effectiveness

Low • • • High

Questions

Assessment

Ranking

1 2 3 4 5

PREPARE AND PRESENT PROPOSAL

TAKE ACTION

You have explored conservation options and decided on which energy-saving plans will be most beneficial in the school setting. Now you can use your powers of persuasion to create a proposal to present to the school committee.

Setting the Stage

Discuss these questions:

1. What conservation problems were discovered during your Energy Audit?
2. How do the energy-saving ideas you chose offer short- and long-term solutions?
3. What main factors made you choose these conservation measures?

Focus

Think about how you will set up your presentation and decide on what tone your proposal should take. What do you want to emphasize? How can charts, graphs, and tables help you illustrate your ideas? What tone do you think will be most persuasive? Work together to create an outline, using the following as a guideline:

I.   Title
II. Introduction
     Include a brief summary of the Energy Audit
     and the investigation your Action Group conducted.
III. Recommendation
     Include information about cost, benefits, step-by-step
     implementation, opportunity for participation,
     maintenance, and so on.
IV. Research and data
     Include facts, figures, projections in the
     form of illustrations, graphs, charts.
V.   Description of conservation
     activities already underway

THINK ABOUT IT

"Building more power plants is like applying higher water pressure to a leaky tub. Conserving our energy is like plugging the drain. More than ever before, we need to conserve our energy and plug our leaks. Fortunately, the leaks aren't hard to find, and neither are the solutions."

from Robbins & Solomon, *Choices for Our Future*, Book Publishing Company, 1994, p. 106

Assign the tasks that need to be done in order to complete the proposal. Use Activity Sheet 15 to keep track of progress as the proposal is prepared.

It's a Wrap

When all the tasks to put the proposal together are complete, review each part and make changes as necessary. Be sure to include data to support your main points. Then present your proposal to the committee.

Home

Write up a proposal for the energy-saving measures your family has agreed to try. Explain how each family member will participate in setting up and carrying out the ideas.

Your teacher will give you a two-part activity sheet
like the one below to use with this lesson.

Name _____

 PROPOSAL CHECKLIST (part 1)

Use this checklist to plan and monitor tasks that may need to be done in order to
complete your proposal. Make a note of who is responsible for completing each task, when
each task should be completed, materials needed, and so on. Add to the list as needed.

TASKS	NOTES
1. TITLE ☐ Cover illustration ☐ Proposal statement	
2. WRITE THE INTRODUCTORY PARAGRAPH. ☐ Explain the project. ☐ Briefly describe audit results.	
3. WRITE RECOMMENDATION. ☐ Outline each plan. ☐ Highlight the benefits. ☐ Specify the costs. ☐ Suggest a step-by-step implementation schedule. ☐ Include ideas for motivating student body, increasing awareness, and encouraging participation. ☐ Outline long-range maintenance requirements, costs, planning. ☐ Pinpoint projected savings.	

Continue recommendations on next page ☞

and explain where to go
from here.

TRACK RESPONSE TO PROPOSAL

You have recommended conservation measures to the school committee, including ways to implement and maintain them. Now you can work to increase awareness and put your proposals into effect.

Setting the Stage

Discuss these questions:

1. How can you find out the effect of your conservation plans?

2. How can you assess energy awareness and participation?

Focus

Discuss the energy-saving proposals that you developed and presented to the committee. How can students at your school be challenged to continue to support the energy-saving plans? How can student participation in conservation measures be increased? Use Activity Sheet 16 to summarize the results of your proposals and to keep track of progress over time.

It's a Wrap

Discuss how the conservation plan is going. What are the most surprising benefits? What would make the plan more effective? Use your Journal to record three ideas for increasing conservation awareness.

Home

Use your Journal to write a progress report about the impact that your conservation plan is having on your household. Explain how it has increased energy conservation awareness and how family members' habits have changed.

THINK ABOUT IT

"Take only what you need and leave the land as you found it."

Arapaho Proverb

from Guy A. Zona, *The Soul Would Have No Rainbow if the Eyes Had No Tears*, Simon & Schuster, 1994, p. 88

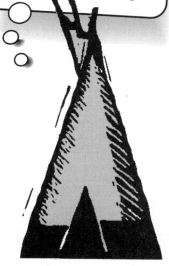

Your teacher will give you a two-part activity sheet
like the one below to use with this lesson.

ACTIVITY SHEET

Name _____

TRACKING SHEET (part 1)

Use this tracking sheet to summarize and monitor the results of your
proposal and to assess students' conservation awareness.

Proposal Summary

Implementation Report

Month 1 Evidence

Month 2 Evidence

Month 3 Evidence

ngs

Issues and Information

ISSUES AND INFORMATION

Although all sources of energy ultimately come from natural processes, nonrenewable resources cannot be replaced naturally at the rate they are used. The fossil fuels—coal, oil, and natural gas—are all examples of nonrenewable resources. It took the earth about a million years to produce the fossil fuels we consume in a single year, and we are using fossil fuels at faster and faster rates. Supplies are running out. In addition, burning fossil fuels has caused extensive damage to the planet. Smog, air pollution, acid rain, global warming, and water pollution are all linked to overuse of fossil fuels. Even if supplies were plentiful, these environmental problems should compel us to find alternative energy resources.

The Fossil Fuels

Oil, coal, and natural gas are called fossil fuels because they are formed from the remains of organisms, mainly plants, that lived millions of years ago. Because fossil fuels are relatively inexpensive and easy to obtain, they have become a common source of energy in many countries. The United States uses fossil fuels to generate most of its electricity. This is an inefficient process; a typical U.S. power company loses about 2.5 times more energy than it delivers. The ready availability of fossil fuels has led to their overuse. And once these nonrenewable fuels have been used up, they cannot be replaced.

Oil Oil, or petroleum, is a liquid fossil fuel. Sometimes oil is found in rock or sand, but generally it is in reservoirs deep underground. Wells are drilled on land or offshore to bring oil to the surface. This crude oil is transported to refineries where it is changed into gasoline and petrochemicals that are used in the manufacture of thousands of different products, including fertilizers, pesticides, plastics, photographic film, ink, medicines. Oil resources are limited, and already more than three-fourths of the oil from the continental U.S. has been extracted and used. At the current rate of use, the U.S. oil supply could be depleted by the year 2020.

Coal Coal is created from decaying plant materials compressed over millions of years to form a solid mass of almost pure carbon. It is the world's most abundant fossil fuel, and much of it is found in the United States. Coal is extracted by surface strip mining and tunnel or pit mining. Strip mining, where the top layer of soil and rock is removed to uncover lower layers of coal, is used to extract more than half of the coal mined in the United States. In many places, strip mining has devastated the land. Tunnel mining can also leave the land unusable because the ground above an abandoned mine can be unstable.

One-third of the energy used throughout the world comes from coal. In the United States, coal accounts for over half of the electricity generated. In addition to supplying fuel, coal is the source of many chemical products, including medicines and pesticides.

Natural Gas Natural gas is a mixture of several gases, predominantly methane, a gas released from rotting plant material. Often found near oil deposits, natural gas is easily extracted by drilling wells. One advantage of natural gas is that it does not require processing, as do coal and oil. After being cleaned of impurities, natural gas is ready for the consumer. While burning natural gas does produce carbon dioxide, it burns hotter and cleaner than other fossil fuels. It is also relatively low priced. For these reasons, natural gas is a choice fuel for residential and industrial heat. Natural gas is also used to generate electricity. While natural gas has considerable advantages, a big problem exists: its supply is limited. According to the U.S. Department of Energy, natural gas reserves may be depleted in only about 60 years.

Environmental Impact of Fossil Fuel Use

The extraction and use of fossil fuels has taken its toll on the environment. Mining, drilling, leaks, and spills damage natural ecosystems, and burning fossil fuels contributes to smog, acid rain, and global warming. In addition, exposure to pollutants poses health risks, including lung problems, skin damage, and some forms of cancer.

Smog and Air Pollution Smog—a mixture of ash, soot, and other pollutants—comes from burning fossil fuels. Smog is especially a problem in densely populated urban areas, where cars and industry rely heavily on fossil fuels. Los Angeles has attracted a lot of attention because of its smog, but many other cities—including Denver, Salt Lake City, Mexico City, and Tokyo—are often blanketed in smog. People who live in polluted areas suffer numerous health problems, including headaches, fatigue, heart and lung disease, and cancer.

Acid Rain Burning fossil fuels produces sulfur dioxide or nitrogen oxide. When these compounds combine with water vapor in the atmosphere, they rain back down to earth as sulfuric or nitric acid, making the rainwater more acidic than vinegar. Coal-burning power plants, which pour 25 million tons of sulfur dioxide into the atmosphere annually, are the primary cause of acid rain. The second major cause is automobile emissions. Acid rain contaminates water systems, killing plant and animal life. It also harms the soil and destroys forests. In Europe, 35% of the forests have suffered damage from acid rain.

Water Pollution Both acid rain and air pollution contribute to water pollution. The process of extracting fossil fuels can also pollute water systems with toxic chemicals. Oil spills can occur when oil is being drilled, pumped, or transported. Every year approximately one ton of every 1000 tons of oil produced ends up polluting the ocean environment. Rivers, lakes, and underground water systems are further polluted with chemicals leached from coal mines. Water polluted with toxic substances destroys ecosystems and jeopardizes human health by contaminating drinking water and water used for recreation.

Global Warming The way in which carbon dioxide and other gases can trap heat in the atmosphere is similar to the way a blanket traps body heat. This natural process, called the greenhouse effect, allows the earth to absorb radiation from the sun. Now, however, there is concern that excessive amounts of carbon dioxide, produced from burning fossil fuels, are causing the earth's temperature to rise. Increased global warming would have profound effects: ocean waters could rise, causing flooding; severe droughts and heavy storms could occur, along with dramatic climate changes; deserts could replace cropland, causing widespread famine. Entire ecosystems would be altered.

Nuclear Power

Nuclear power is produced in a process called nuclear fission, in which atoms are split and the neutrons within the atoms are set free. These neutrons start a chain reaction that produces tremendous energy. The chain reaction requires the use of uranium, a radioactive element. Nuclear reactors (nuclear power plants) contain and control this chain reaction, releasing heat at a controlled rate. Nuclear reactors are expensive to build, but they can make plentiful supplies of fuel without producing carbon dioxide. About 430 nuclear reactors have been built worldwide in countries such as Japan, India, France, Great Britain, and the United States.

Safety in terms of human health and the environment is a key issue in the use of nuclear power. Nuclear reactors can leak radiation, causing at the very least a serious health hazard for workers. Accidents can occur because of human or mechanical errors, as was the case at Three Mile Island near Harrisburg, Pennsylvania in 1979. Safety standards imposed by the Nuclear Regulatory Commission in response to the accident have made nuclear power plants too expensive to construct and maintain. In 1986, an explosion and fire occurred in one of four nuclear reactors located at Chernobyl in the former Soviet Union, releasing enormous quantities of radioactive steam. Thirty-one people died from the blast and 135,000 from within a 1000-mile radius had to be evacuated. Radiation released into the atmosphere contaminated crops in Europe and was carried by air currents around the world; chronic health problems associated with exposure to radiation as a result of this accident are not yet fully realized.

Even if nuclear reactors are securely built, nuclear fission naturally produces dangerous radioactive wastes. It takes thousands of years for this material to lose its radioactivity. In the meantime we have no way to safely dispose of it.

During the early 1970s, nuclear power looked promising as an alternative to fossil fuels. Since the 1980s, however, public support for nuclear power plants has steadily decreased in the United States and in other countries, mainly because of health, environmental, safety, and economic concerns.

ISSUES AND INFORMATION

Section B
RENEWABLE & ALTERNATIVE ENERGY RESOURCES

Conservation can help to extend the supply of precious nonrenewable resources, giving us time to develop other energy resources that can free us from our dependence on nonrenewable fossil fuels. Many alternative energy resources are now available, but they are not widely used. Getting people to switch to alternative energy resources depends on making the technologies for using them affordable, providing economic incentives for research and development, educating consumers, and gathering political support for their use.

A renewable energy resource is constantly renewed by natural processes. Carefully managed, alternative energy from renewable resources can last indefinitely. Renewable energy resources include heat from the sun, from the earth's molten core, and from burning plants (biomass). Resources related to solar energy include power from wind, falling water, and tides and ocean currents.

The U.S. Department of Energy estimates that America could get up to 70% of its total energy from sun, wind, water, geothermal, and biomass resources within the next 40 years. The challenge is to manage these renewable resources in ways that are both economically and environmentally sensible.

Biomass and Biogas

Biomass Biomass is organic material produced by plants and animals and burned for fuel. The most common example is firewood; other examples are sawdust, cow droppings, peat, and organic garbage. Biomass is the main source of energy for nearly half the world. Half of all wood cut worldwide each year is used as cooking and heating fuel.

Biomass has had the advantage of generally being inexpensive and widely available. However, it has a number of disadvantages today. Burning biomass pollutes the environment with carbon dioxide, smoke, and ash. Also, in many parts of the world (including the rain forests), cutting and burning trees and shrubs for fuel is taking place at a much faster rate than the plants can be replenished. With a growing world population, more than 1.5 billion people every day have difficulty finding enough wood for fuel. Harvesting so much wood for fuel is contributing to deforestation that is destructive not only to the local ecology but also to the global environment.

New methods are being used to convert biomass into a cleaner burning gas or liquid. The gas fuel ethanol, made from corn or sugar cane, is used to fuel about half of the cars in Brazil. The large quantity of plant material required to make this fuel poses a problem because it requires land that could be used for food production or left uncultivated for environmental reasons.

Biogas Methane is a natural gas that results when organic material decomposes through bacterial activity in an airtight environment, as in a landfill. Biogas generators convert plant and animal waste to methane. China currently uses over 7 million biogas generators, and India has thousands in operation. In the United States methane gas is produced in landfills, although less than 100 of the 3000 or so landfills are equipped with biogas generators. Biogas generators can also capture methane produced from manure on feedlots and farms. Some farms power generators and heaters with the methane they produce. Biogas opponents feel that organic plant waste should be composted and that manure should be used for fertilizer.

Geothermal Power

Geothermal ("heat from the earth") energy is produced when steam or hot water from an underground source is piped to turbines, which then produce electricity. Underground masses of hot rock, volcanoes, and geysers are sources of geothermal energy. A number of countries tap this renewable energy resource. In Italy, geothermal energy powers the country's railway system. Iceland, a site of high volcanic activity, has an abundance of geothermal power that is used to heat the majority of its homes. The geyser field stream north of San Francisco has been producing electricity since 1969 and by 1987 was supplying more than 2% of California's electricity at half the cost of a new nuclear plant.

Hydroelectric Power/ Hydropower

Hydropower ("water" power) is generated when water is put under tremendous pressure and then allowed to flow with great force. Hydropower is most commonly produced by damming rivers. The water is backed up and then released to fall over the dam. The force of the falling water causes turbines to rotate and send the water's mechanical energy into a generator where electricity is produced. Hydropower is generated throughout the world, especially in countries with mountainous areas, such as Norway, Canada, Austria, and Switzerland. Hydropower supplies Norway with all of its electricity. The largest hydroelectric plant in the world, the Itaipu Dam situated between Brazil and Paraguay, generates 12,600 megawatts of power every year.

Although hydroelectric power is renewable and non polluting, dams can cause considerable environmental damage by disrupting a river's natural flow and the delicate ecosystems that depend on the river. These problems can be avoided if the force of naturally occurring waterfalls is used to generate hydroelectric power. Niagara Falls, between Lake Erie and Lake Ontario, is a source of electricity for both Canada and the United States.

Hydrogen

Hydrogen is a simple, readily obtainable gas. It can be used to generate electricity, power industry, run home appliances, and fuel automobiles, jet planes, and spaceships—all without polluting the atmosphere. It is relatively easy to redesign a car engine so that it operates on hydrogen. Hydrogen-fueled cars are already operating in Japan, and hydrogen-fueled buses are used in Germany. Storage of the fuel in the vehicle can be inconvenient; it is typically stored in a bulky, heavy pressurized tank mounted in the trunk. Also hydrogen filling stations do not exist, so it takes a long time to fill the pressurized tank. Hydrogen fuel today is considerably more costly than conventional fuels, but technological advances promise to make conversion to hydrogen fuel more economical and efficient.

Ocean Power

Ocean tides, caused by the gravitational pull of the moon and sun, can be used to generate power. In a location where tides rise higher than average, a dam can be built across a bay or inlet. Water pours into the dam at high tide and turns turbines that generate electricity. Tidal power systems are at work in a number of countries, including Norway, China, the United States, and Canada. The most successful tidal power system in the world is in Brittany, France, where a dam produces power for a community of over 46,000 people. Tidal systems alone may not be sufficient for large communities and backup systems may be required. In addition, care must be taken to protect coastal habitats that may be affected if tidal action is altered.

Solar Energy

Every day, solar radiation strikes the earth in amounts equal to all the energy from 173 million large power stations. Some of this energy is reflected back into space, and some is used up in the water cycle that produces rainfall. The remaining energy represents a limitless source of clean fuel. Two main methods for collecting and using solar energy are solar thermal and photovoltaics.

Solar Thermal On a small scale, solar thermal energy can be collected by a relatively simple system that operates passively or actively. South-facing, double-glazed windows are a popular and relatively inexpensive way to passively collect solar energy in a home or building. In an active solar energy system, a set of solar collectors is mounted on a roof with a southern exposure and connected to an electric pump and fan. Active solar systems can supply energy for heating space and water. Solar water heating is widely used in some areas. In Israel, for example, 65 percent of all domestic hot water is heated using solar energy. In the United States, over 1.3 million active solar hot water systems are operating, especially in California, Florida, and the Southwest. The technology for solar heating systems is readily available, although the initial cost of installing an active solar system can be more costly than other conventional heating systems. In cooler climates, solar collecting systems require a backup system.

On a much larger scale, complex solar collectors consisting of reflecting mirrors concentrate the sun's rays to heat liquids to as high as 3000° Celsius. Such systems require large stretches of land. In the Mojave Desert of California, for instance, more than a million mirrors are set on 400 acres to concentrate solar heat on long tubes filled with oil. The heat is used to turn a turbine and generate electricity for Los Angeles.

Photovoltaics Also called solar cells or PVs, photovoltaics contain elements such as silicon that react to sunlight by releasing electrons. This chemical change produces an electrical current that is then transferred through wires to storage batteries or machines. When many solar cells are required, they are collected together on panels. Photovoltaics have been improved since their introduction in the 1950s and are now made with new materials so that they can operate even on cloudy days. Solar-powered calculators and watches and R.V. battery chargers use photovoltaics. In the United States, photovoltaics are also used for lighting street signs and lighthouses. Expanded use depends on further technological breakthroughs that will reduce their cost and increase their efficiency.

Wind
····

For thousands of years, people have used windmills to convert wind energy into mechanical power. Early American settlers used windmills for various tasks, such as sawing wood, pumping water, and grinding grain. In the early twentieth century, windmills were used to produce electricity, but as electrical power lines became more available, the use of windmills diminished. A renewed interest in wind power came with the oil shortages in the 1970s. Huge wind turbines were connected to electricity generators that converted wind power into electricity. Wind turbines erected back then were noisy and unstable; modern wind generators run more efficiently.

Wind farms, built in various windy parts of the world on large stretches of land, have been able to provide power efficiently without polluting the atmosphere. By 1989, wind farms in California contributed over 1 percent of the state's total electrical power at a cost far below that of nuclear power. Smaller personal wind turbines often found on farms have also been successful. Although they do not pollute the atmosphere, wind farms can impact the environment by harming wildlife.

MEASURING ENERGY USE AND COSTS

The following equivalents and formulas will help you take measurements and calculate costs.

Units of Measurement

Length

Metric

1 kilometer (km) = 1000 meters (m)
1 meter (m) = 100 centimeters (cm)
1 meter (m) = 1000 millimeters (mm)

English

1 foot (ft) = 12 inches (in)
1 yard (yd) = 3 feet (ft)
1 mile (mi) = 5280 feet (ft)

Metric-English

1 kilometer (km) = 0.621 mile (mi)
1 meter (m) = 39.4 inches (in)
1 inch (in) = 2.54 centimeters (cm)
1 foot (12 in) = .305 meter (m)
1 yard (3 ft) = 0.914 (m)

Area

Metric

1 square kilometer (km^2) = 1,000,000 square meters (m^2)
1 square meter (m^2) = 1,000,000 square millimeters (mm^2)
1 hectare (ha) = 10,000 square meters (m^2)
1 hectare (ha) = 0.01 square kilometers (km^2)

English

1 square foot (ft^2) = 144 square inches (in^2)
1 square yard (yd^2) = 9 square feet (ft^2)
1 square mile (mi^2) = 27,880,000 square feet (ft^2)
1 acre (ac) = 43,560 square feet

Metric-English

    1 hectare (ha) = 2.471 acres (ac)

    1 square kilometer (km²) = 0.386 square mile (mi²)

    1 square meter (m²) = 1.196 square yards (yd²)

    1 square centimeter (cm²) = 0.155 square inch (in²)

Volume

Metric

    1 liter (L) = 1000 milliliters (mL) = 1000 cubic centimeters (cm³)

    1 milliliter (mL) = 0.001 liter (L)

    1 milliliter (mL) = 1 cubic centimeter (cm³)

English

    1 gallon (gal) = 4 quarts (qt)

    1 quart (qt) = 2 pints (pt)

    1 pint (pt) = 2 cups (c)

Metric-English

    1 liter (L) = 0.265 gallon (gal)

    1 liter (L) = 1.06 quarts (qt)

    1 liter (L) = 0.0353 cubic foot (ft³)

    1 cubic meter (m³) = 35.3 cubic feet (ft³)

    1 cubic kilometer (km³) = 0.24 cubic mile (mi³)

    1 barrel (bbl) = 159 liters (L)

    1 barrel (bbl) = 42 U.S. gallons (gal)

Mass

Metric

    1 kilogram (kg) = 1000 grams (g)

    1 gram (g) = 1000 milligrams (mg)

    1 metric ton (mt) = 1000 kilograms (kg)

English

    1 ton (t) = 2000 pounds (lb)

    1 pound (lb) = 16 ounces (oz)

Metric-English

    1 metric ton (mt) = 2200 pounds (lb) = 1.1 tons

    1 kilogram (kg) = 2.20 pounds (lb)

    1 pound (lb) = 454 gram (g)

    1 gram (g) = 0.035 ounce (oz)

Temperature

Fahrenheit (°F) to Celsius (°C): °C= (°F-32.0)/1.80
Celsius (°C) to Fahrenheit (°F): °F= (°C × 1.80) + 32.0

Energy and Power

Metric
1 kilojoule (kJ) = 1000 joules (J)
1 kilocalorie (kcal) =1000 calories (cal)
1 calorie (cal) = 4.184 joules (J)

Metric-English
1 kilojoule (kJ) = 0.949 British thermal unit (Btu)
1 kilojoule (kJ) = 0.000278 kilowatt-hour (kWh)
1 kilocalorie (kcal) = 3.97 British thermal units (Btu)
1 kilowatt-hour (kWh) = 860 kilocalories (kcal)
1 kilowatt-hour (kWh) = 3400 British thermal units (Btu)
1 quad (Q) = 1,050,000,000,000,000 kilojoules (kJ)
1 quad (Q) = 2,930,000,000,000 kilowatt-hours (kWh)

Approximate crude oil equivalent
1 barrel (bbl) crude oil = 6,000,000 kilojoules (kJ)
1 barrel (bbl) crude oil = 2,000,000 kilocalories (kcal)
1 barrel (bbl) crude oil = 6,000,000 British thermal units (Btu)
1 barrel (bbl) crude oil = 2000 kilowatt-hours (kWh)
1 gallon oil = 140,000 British thermal units (Btu)

Approximate natural gas equivalent
1 cubic foot (ft³) natural gas = 1000 kilojoules (kJ)
1 cubic foot (ft³) natural gas = 260 kilocalories (kcal)
1 cubic foot (ft³) natural gas = 1000 British thermal units (Btu)
1 cubic foot (ft³) natural gas = 0.3 kilowatt-hours (kWh)

Approximate hard coal equivalent
1 ton (t) coal = 20,000,000 kilojoules (kJ)
1 ton (t) coal = 6,000,000 kilocalories (kcal)
1 ton (t) coal = 20,000,000 British thermal units (Btu)
1 ton (t) coal = 6000 kilowatt-hours (kWh)

Approximate propane equivalent
1 gallon propane = 93,500 British thermal units (Btu)

Energy and Power Calculations

Energy can be measured in the following units:
- British Thermal Units (Btu's)
- Calories
- Joules
- Therms
- Watts, Kilowatts, and Kilowatt-hours

Here are some important conversions:
- 1055 Joules = 1 Btu
- 252 calories = 1 Btu
- 1 kilowatt-hour of electricity = 3413 Btu's
- 1 cubic foot of natural gas = 1030 Btu's
- 1 therm = 100,000 Btu's

British Thermal Units (Btu's)

The British thermal unit or Btu is the most common unit for measuring all forms of energy. It is used to measure electricity, oil, and gas. One Btu is a very small amount of energy. It is the amount of heat required to raise the temperature of one pound of water one degree Fahrenheit. Btu/hour ratings are shown on labels on gas appliances.

Calorie

A calorie is the unit of heat required to raise the temperature of 1 gram of water by 1 degree Celsius. Calories are also used to measure the amount of energy-producing potential contained in foods.

Joules

A Joule is a measure of electrical energy or physical energy. Power, expressed in Joules, is the amount of energy used divided by time. A night light, for example, uses about 5 Joules per second.

Therms

A therm is a unit of heat. For example, natural gas from different sources can produce different amounts of heat; therms are used to measure the energy content of the gas. A therm is equivalent to 100,000 Btu's.

Watts, Kilowatts, and Kilowatt-hours

Watts are the unit of power, or the rate, at which electrical work is done. Because a watt is a small unit, kilowatts are more frequently used. A kilowatt equals 1000 watts. The amount of electricity we use is measured in kilowatt-hours (kWh). One kilowatt-hour equals the amount of electricity needed to power one kilowatt for one hour. Ten 100-watt light bulbs used for an hour would use one kWh, as would a 1000-watt heater used for one hour. Electric appliances have labels on the back giving their wattage rating.

Calculating Energy Use and Costs

Electric Appliances

1. Change watts to kilowatts by dividing number of watts by 1000.

   kilowatts = watts ÷ 1000

   Example: CD player (100 watts) = .1 kilowatts

2. Calculate kilowatt-hours (kWh) by multiplying number of kilowatts by the number of hours in use.

   kilowatt-hours = kilowatts × hours used

   Example: To operate a CD player for 4 hours, multiply .1 kW by 4.

   .1 × 4 = .4 kWh

3. To calculate cost, obtain the current cost of electricity per kWh from an electric bill or from your utility company. (You may have to divide the total cost on the bill by the number of kWh used to get the rate per kWh.) Multiply the kWh by the current cost of electricity per kWh.

   $$cost = kWh \times cost/kWh$$

   Example: A CD player operated for 4 hours uses .4 kWh. At a rate of $0.09/kWh:

   $$.4 \times \$0.09 = \$.036$$
   It costs about 4 cents.

Gas Appliances

1. Multiply the appliance's Btu/hour rating by the number of hours used and divide by 100,000 to calculate the therms used.

$$\text{therms} = \frac{\text{Btu/hour} \times \text{hours}}{100{,}000}$$

2. To calculate the cost, obtain the current cost of gas per therm from a gas bill or from your local gas company. (You may have to divide the total cost by the total gas use in therms to get the rate per therm.) Multiply the therms used by the current cost of gas per them.

$$\text{cost} = \text{therms} \times \text{cost/therm}$$

READING UTILITY BILLS

ISSUES AND INFORMATION

The following information will help you use the information on utility bills.

TERMS FREQUENTLY FOUND ON FUEL BILLS

Electric Bills

EST	Estimated Meter Reading
kW	Kilowatt
kWh	Kilowatt-hour, a unit of electric energy
Basic/Service Charge	A monthly service fee for maintenance and other services, charged whether or not a customer uses any electrical power
Baseline Charge	The lowest charge a customer may receive for the service; an energy-use allowance calculated on the customer's climate zone
CR	Credit Balance
Demand Charge	A charge of maximum power measured in kilowatts during the billing period
Power Factor or Energy Cost Adjustments	Adjustments reflecting variations in the costs of fuel or energy

Gas Bills

CCF	Hundreds of cubic feet. Gas meters record usage in CCF.
Btu/Billing Factor	The average heating value of the gas for the billing period. It is used to convert CCF to therms.

Therm	A standard unit of heat energy. A therm is 100,000 Btu's. To determine the number of therms used, multiply the CCF by the Btu factor.
kWh	Kilowatt-hour, a unit of electric energy. One therm equals 29.307 kWh.
Baseline	The amount of gas billed at the lowest residential rate.
Utility User's Tax	A tax charge by some cities and counties based on the amount of the current monthly gas bill.

Note that both electric and gas charges may include two price rates: One is a "baseline usage" rate or "base minimum charge"; the other is "over baseline usage" rate. These may also be called "summer rate" and "winter rate." The baseline usage rate is the lowest charge and is an energy-use allowance calculated for the customers' climate zone.

Oil Bills

Oil utility companies charge for gallons of oil used. The bills usually show
- Number of gallons used
- Rate per gallon
- Total bill (gallons used x rate)

Sometimes the gallons of oil used will be converted to Btu's:
    1 gallon oil = 137,797 Btu's

CONSERVATION MEASURES

ISSUES AND INFORMATION

The success of any conservation measure depends on three actions: change personal behavior, increase efficiency, and reduce waste. Below you will find examples of each of these actions and suggestions for how they can be applied to common energy-using systems and equipment.

All Equipment and Systems

Change Personal Behavior

- Consider carefully whether you really need a new appliance.
- Limit your use of unnecessary appliances such as electric toothbrushes, can openers, and so on.
- Turn off all appliances (TVs, VCRs, computers, stereo equipment, office equipment) when not in use.
- Make a nighttime Put-to-Bed checklist: reduce or turn off lights; turn off equipment; adjust thermostats.

Increase and Maintain Efficiency

- When you do purchase appliances, choose the most energy efficient ones; look for and study the EnergyGuide labels.
- Choose appliances that use alternative sources of energy whenever possible (for example, solar calculators).
- Keep all appliances in good working order.

Reduce Wasted Energy

- Turn off or adjust furnaces, pilot lights, cooling systems, water coolers, and all equipment when not in use for extended periods of time (such as holidays, vacations).

Space Heating

Change Personal Behavior

- Keep heat in by closing curtains on cold days and at night.
- Be sure windows are closed in heated rooms.

- Reduce your living area at home in the winter to only a few rooms and heat only those spaces (use space heaters).
- Close off heat to rooms or areas not in use (storage areas, entry ways, and so on).
- Wear warmer clothing and use blankets instead of more heat; a sweater may be all that is needed!
- Close the damper when the fireplace is not in use.

Increase and Maintain Efficiency

- Set the thermostat at 68°F during the day or lower and 55°F or lower at night.
- Choose highly energy-efficient heating systems.
- Use a programmable thermostat (with an automatic timer) to lower the heat at night.
- Keep furnace in good repair; clean or replace filters once or twice a year or as often as manufacturer instructs.
- Maximize passive solar energy to heat rooms on sunny days.

Reduce Wasted Energy

- Insulate walls, ceilings, and attics.
- Caulk and weather-strip doors and windows; caulk any other air leaks around pipes and other openings; patch holes even in internal walls.
- Seal heating ducts (with duct tape or caulking) from leakage at joints, elbows, and connections.
- Add storm windows (plastic or glass) to single-glazed windows; replace single-glazed windows with double-glazed or other energy-efficient models.
- Use insulated window and door coverings.
- Seal off electrical outlets and switch boxes with foam gaskets or insulation.
- Repair broken windows.
- Install storm doors to protect entrances from drafts.
- Turn the gas furnace pilot light off in warm seasons or when away for extended periods of time (such as vacations).
- Contact your local utility company and ask about obtaining an energy audit to evaluate energy losses.

Water Heating

Change Personal Behavior
- Take shorter showers.
- Don't leave water running while brushing teeth, shaving, or washing.

Increase and Maintain Efficiency
- Choose an energy-efficient water heater (check Energy Guide labels).
- Insulate water heater with a blanket and wrap insulation around water pipes.
- Lower the water heater thermostat to 120°F or less.
- Repair leaky faucets.
- Install low-flow shower heads and install aerators on faucets.

Reduce Wasted Energy
- Set back or turn off the water heater temperature when you go away for an extended time (such as holidays, vacations).

Cooling

Change Personal Behavior
- Limit the use of stove and oven during the hottest part of the day.
- Be sure windows are closed in cooled rooms.
- When using hot water in the bathroom, close the door and open windows to allow heat and moisture to escape outside.
- Do heat-intensive activities, such as clothes-drying, during the cool times of the day.

Increase and Maintain Efficiency
- Choose an energy-efficient air conditioning system or, better yet, use fans (free standing, window, ceiling, or attic).
- Set air conditioning at 78°F or higher in summer.
- Install a programmable thermostat to use less cooling at night (raise the temperature).
- Seal cooling ducts from leakage at joints, elbows, or connections (use duct tape or caulking).

Reduce Wasted Energy

- Cool only those rooms that are in use (not storage areas, entry ways, and so on).
- Maximize cool air and minimize heating from the sun by closing windows and drapes in the morning, then opening them at night.
- Use shading devices, such as awnings, to block sun from south-facing windows and from the air-conditioning condenser.
- Shade the west, south, and east sides with trees and vines. Plant deciduous trees on south and southwest sides to shade in summer and allow sun in winter. Plant evergreens on the north and north-west sides to shade in summer and break the wind in winter.

Lighting

Change Personal Behavior

- Turn off unneeded lights (especially when leaving a room).
- Turn on only the lights you need for safety and work.

Increase and Maintain Efficiency

- Use natural lighting whenever possible ("daylighting").
- Use fluorescent and other energy-saving bulbs wherever possible.
- When you buy bulbs, buy ones with the needed amount of lumens rather than watts.
- Clean lamps, tubes, and light fixtures regularly.
- Install automatic (programmable) lighting controls to turn lights off when areas are not in use.

Reduce Wasted Energy

- Reduce lighting levels by using dimmers and lower wattage.
- Install motion sensors where possible to avoid unnecessary continuous lighting.
- If a burnt-out bulb hasn't been missed, don't replace it unless it causes an inconvenience or a safety hazard.

Computer and Office Equipment

Change Personal Behavior

- Turn off computer systems, copy machines, and printers if they are not in use for more than an hour (screen savers do not save energy, they only save phosphors).
- Use double-sided copy feature whenever possible.
- Recycle used paper.

Increase and Maintain Efficiency

- Buy computer equipment with the "Energy Star" label.
- Install plug-in timers to automatically turn off machines at night and on weekends.

Reduce Wasted Energy

- Be sure energy-saver features are operating on copy machines.

Kitchen Appliances

Change Personal Behavior

- Avoid letting the refrigerator door stand open; decide what you want before opening the door.

Increase and Maintain Efficiency

- Vacuum or dust the refrigerator condenser coils at least once a year (the copper coils in back).
- If the refrigerator is several years old, install a GreenPlug (sold in hardware/home improvement stores) to save electricity by reducing the voltage feeding the refrigerator.
- Defrost before ice builds up to 1/4-inch thick.

Reduce Wasted Energy

- Do not place the refrigerator or freezer next to a stove, oven, or heating duct.

Cooking

Change Personal Behavior
- Preheat the oven only as long as necessary.
- When cooking small meals, use a microwave or toaster oven.
- Instead of defrosting in an oven or microwave, let frozen food thaw in the refrigerator or on a counter.

Increase and Maintain Efficiency
- Add an automatic ignition to improve efficiency of gas stoves.

Reduce Wasted Energy
- Use cooktop burners rather than the oven.

Dishwashing

Change Personal Behavior
- Wash dishes by hand using plastic pans for soap and rinse water.

Increase and Maintain Efficiency
- Wash only full loads and use the shortest cycle that will clean the dishes; some models have a water miser option.
- If possible, turn the dishwasher off before the drying cycle and allow the dishes to air dry; some models can be set to air dry.

Reduce Wasted Energy
- Select the air-dry and water-miser settings on the dishwasher.

Clothes Washing and Drying

Change Personal Behavior
- Use cold and cool water washes and rinses for clothes when possible.
- When possible, hang your clothes to dry on a clothesline.

Increase and Maintain Efficiency
- Clean the lint filter before each load in a dryer.

Reduce Wasted Energy
- Use the washer and dryer only when you have enough clothes for a full load.

Transportation

Although you have been focusing on energy uses in your school and your home in this project, transportation is one of the biggest energy-users in the world. Here are some tips to help save energy as you're getting around.

- Drive less; use public transportation, ride a bicycle, and walk whenever possible.
- Carpool to school, work, and community events.
- Drive cars that use less gas per mile; good mileage should be a primary consideration when purchasing a car.
- Maintain cars in good working order, including tire pressure.
- Drive at the speed limit; avoid quick acceleration and braking.
- Organize errands and trips to minimize extra driving.
- Remove any extra weight in your car.

Recycling

Recycling or reusing products is an effective conservation measure that depends on changing behavior and taking personal responsibility. Become an informed consumer by paying attention to labeling and materials before you buy. Purchasing products manufactured with post-consumer waste instead of new materials and recycling or reusing paper, plastics, glass, and yard waste will go a long way toward improving the environment, preserving resources, cutting costs, and saving energy, too.

Section F
ENERGY USE OF APPLIANCES

The following figures for appliance energy use are approximate.

Electric Appliances	Energy Use in Watts
Air Conditioner	
Room	1500/Hour
Central, 5-Ton Unit	7500/Hour
Electric Blanket	75/Night
Blender	350/Hour
Clock	2/Hour
Clock Radio	6/Hour
Clothes Dryer	5000/Load
Clothes Washer	250/Load
Coffee Maker	120/Use
Computer	75/Hour
Dishwasher	
Normal Cycle	1000/Load
Energy Saver	500/Load
Fan (Portable)	250/Hour
Freezer (20 Cu. Ft.)	3800/Day
Garbage Disposal	450/Use
Hair Dryer	100/Use
Heating	
Portable Heater	1500/Hour
Central Heater	25,000/Hour
Blower for Gas Furnace	350/Hour
Iron	1000/Hour

Lighting	
Incandescent 100-Watt	100/Hour
Fluorescent 40-Watt	50/Hour
Microwave Oven	1000/Hour
Mixer	125/Hour
Oven	4000/Use
Refrigerator (22 Cu. Ft.)	5000/Day
Stereo Components	
Cassette Player with Receiver	100/Hour
CD Player with Receiver	100/Hour
Receiver/Radio	75/Hour
Television (Color)	230/Hour
Toaster	75/Use
Toaster Oven	500/Hour
Vacuum Cleaner	750/Hour
VCR	25/Hour

Gas Appliances	Energy Use in Btu
Clothes Dryer	22,000/Hour
Range	30,000/Hour
Outside Grill	35,000/Hour
Water Heater	40,000/Hour

AVERAGE ENERGY USE
IN HOMES AND SCHOOLS

Average Residential Energy Use in California*

Space heating and cooling	30%
Water heating	26%
Refrigeration	11%
Cooking	5%
Clothes drying	4%
Motors	4%
Color TV	2%
Miscellaneous	18%

Average School Energy Use in California*

Space heating and cooling	49%
Lighting	36%
Water heating	7%
Miscellaneous	8%

* Source: *California Energy Demand 1989–2009,* California Energy Commission

Section G
ENERGY EFFICIENCY MEASURES

Use the following to help you evaluate conservation measures that can be used at home or at school and make informed decisions about steps that can be taken to cut energy use. You don't need to be an engineer, you don't need to start over from the ground up, and you don't necessarily need to spend a lot of money to put many of these conservation ideas in motion.

Heating and Cooling

Caulking and Weather-stripping Plenty of energy escapes through the cracks. Sealing doors and window joints with caulk or weather-stripping is an effective and inexpensive measure to conserve cool air in summer and hot air in winter.

Insulation Putting on an extra layer of clothing will keep you warmer on a chilly day. In the same way, insulation in the attic, walls, and floor can help keep a building warm on the inside and slow down heat loss. The R-value of insulation indicates its ability to resist heat flow—the higher the better. Insulation choices include fiberglass and cellulose. Cellulose is environmentally friendly, nontoxic, noncarcinogenic; it is also more economical and energy efficient. Fiberglass contains blown glass and formaldehyde, which can cause skin and respiratory ailments and may present additional health concerns. And when you are insulating, don't forget your water heater and hot-water pipes to slow down heat loss.

Windows Windows let in light and give you a view of the outside world, and if they are single-glazed, you can almost see the heat leaking out. Storm windows can be added to single-glazed windows and may be anything from a simple layer of plastic to Plexiglas or sealed panes of glass to add another layer between you and the outside world. Newer, energy-efficient windows come double glazed or even triple glazed to begin with and may be required by building codes in some climates.

Thermostat Controls Thermostat controls can be used intelligently to keep building temperatures within a comfortable range. Automatic settings can be programmed to kick on only when necessary, on when rooms are in use and off when they are unoccupied or when the outside temperature rises. Central controls are not so efficient as separate controls for rooms or areas, so that unused parts of the building are not heated unnecessarily. Radiators, boilers, and water heaters are all regulated by thermostats.

Landscaping Paving, walkways, and parking lots may be convenient, but they can create a heat trap surrounding a building, raising the air temperature so that additional air conditioning is needed to cool things down. Cool down sunny facades with deciduous trees that will provide leafy shade in hot months and will let the sun in during winter. Planting low-maintenance ground covers and shrubs around buildings and deciduous vines on building walls can also cool outside temperatures through the process of transpiration: Plants absorb water from the soil and release it through their leaves.

Lighting

One of the main things to remember about lights is to turn them off. And when you do use lights, choose bulbs that burn efficiently. A fluorescent bulb uses the same amount of wattage to produce four times as many lumens as an incandescent bulb. Compact fluorescent bulbs are the energy-conscious choice for light fixtures that get a lot of use. They cost more, but they last about 10 times longer.

Motion, heat, or noise sensors can be installed so that lights automatically go on when a room is occupied and off when it is empty.

Outdoor lighting can be improved and made more efficient by retrofitting incandescent or mercury vapor lights with high-density discharge (HID) lamps. Examples are high-pressure sodium (HPS) lamps or metal halide lamps.

Don't forget to flick the switch on your way out, and as you exit, think about this: If EXIT signs were upgraded to use energy-efficient fluorescent or LEDs (light-emitting diodes), an estimated 2.7 billion kWhs of electricity would be saved each year, equal to planting 780,000 trees, losing 380,000 cars, pumping 260 million fewer gallons of gas.

Appliances

Keep It Clean Maintaining appliances already in place will help increase energy efficiency. Clean or replace filters, clean air vents, dust refrigerator coils, replace hoses, defrost freezer compartments, lubricate moving parts according to manufacturer's guidelines.

Read the Fine Print If you are choosing a new appliance, such as a refrigerator, water heater, or washing machine, look at the Energy Guide label to determine the model that meets your needs and saves the most energy, too. The label displays the estimated cost to run the appliance for a year and compares the amount to comparable models. Air conditioner Energy Guide labels rate the energy efficiency of the model and chart the yearly operating costs comparable to other models; heating units have a standard label indicating standards required by regulations, and the manufacturer supplies a more detailed analysis.

You can save substantially on home heating and cooling energy costs by following the simple steps outlined below:

1. Weatherproof your house
2. Assure energy efficient heating and cooling equipment selection and installation
3. Operate and maintain your system to conserve energy.

Help conserve energy. Compare the energy efficiency rating and cost information for this model with others Check the figures and spend less on energy.

Your contractor has the energy fact sheets. Ask for them.

Important Removal of this label before consumer purchase is a violation of federal law (42 U.S.C. 6302)

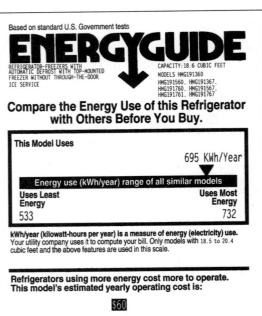

Based on standard U.S. Government tests

REFRIGERATOR-FREEZERS WITH AUTOMATIC DEFROST WITH TOP-MOUNTED FREEZER WITHOUT THROUGH-THE-DOOR ICE SERVICE

CAPACITY:18.6 CUBIC FEET
MODELS HMG191360
HMG191560, HMG191367,
HMG191760, HMG191567,
HMG191761, HMG191767

Compare the Energy Use of this Refrigerator with Others Before You Buy.

This Model Uses

695 KWh/Year

Energy use (kWh/year) range of all similar models

Uses Least Energy
533

Uses Most Energy
732

kWh/year (kilowatt-hours per year) is a measure of energy (electricity) use. Your utility company uses it to compute your bill. Only models with 18.5 to 20.4 cubic feet and the above features are used in this scale.

Refrigerators using more energy cost more to operate. This model's estimated yearly operating cost is:

$60

Based on a1996 U.S. Government national average cost of 8.67¢ per kWh for electricity. Your actual operating cost will vary depending on your local utility rates and your use of the product.

Crunch the Numbers Calculate the cost of running an appliance over its estimated lifetime using the following formula: Purchase Price + Lifetime Energy Cost = Total Cost. The appliance that costs more to purchase but uses less energy per year will wind up costing you less over its lifetime, and well-maintained appliances can last far longer than their average life expectancy.

GLOSSARY

acid rain The burning of fossil fuels emits sulfur and nitrogen oxides into the air. These pollutants combine with moisture in the atmosphere to form sulfuric and nitric acids that fall as rain (or snow or fog). Acid rain can dissolve stone, change the pH balance of rivers and lakes, damage vegetation, and harm wildlife.

biogas Energy generated by burning methane gas. Methane is a gas produced from the decomposition of biomass. It is also the main component of natural gas, a fossil fuel.

biomass Organic material from plants; an estimate of the total mass of organisms that make up all or part of a population, community, or other specified unit. Biomass can be burned to produce heat energy or can be converted into liquid and gaseous biofuels.

British thermal unit (Btu) The quantity of heat needed to raise the temperature of 1 pound of water by 1 degree F.

caulking To make watertight or airtight by filling or sealing with caulk.

coal A solid, fossil fuel found in layers beneath the surface of the earth. Coal is combustible. It usually contains from 40% to 98% carbon mixed with varying amounts of water (2% to 50%) and small amounts of nitrogen (0.2% to 1.2%) and sulfur (0.6% to 4%) compounds. It is mined and used primarily as a fuel to generate steam for the production of electricity. Coal is graded on the basis of heat content and classified as anthracite, bituminous, sub bituminous, or lignite.

crude oil A gooey liquid mixture of hydrocarbon compounds (90% to 95% of its weight) and small quantities of compounds containing oxygen, sulfur, and nitrogen. It is extracted from underground deposits and then sent to refineries to be converted into useful materials such as heating oil, diesel fuel, gasoline, and tar.

deforestation To cut down and clear away trees from a forested area. Although wood is a renewable resource, deforestation depletes this resource faster than it can be replaced and has a devastating effect on the environment and wildlife habitats.

fluorescent light Light and heat created by electricity that excites a gas in the light bulb. Traditional fluorescent bulbs are long tubes, but compact fluorescent bulbs can be used in ordinary light sockets and are energy efficient.

fossil fuels Deposits of decayed plants and animals that have been converted to crude oil, coal, natural gas, or heavy oils by exposure to heat and pressure in the earth's crust over hundreds of millions of years.

geothermal energy Heat transferred from the earth's intensely hot molten core to underground deposits of dry steam (steam with no water droplets), wet steam (a mixture of steam and water droplets), hot water, or rocks lying relatively close to the earth's surface. Geothermal energy is not strictly renewable, since heat reservoirs accessible from the earth's surface can be depleted. However, the resource is extremely large. Hot springs are an example of geothermal activity.

glaze Coat, fit, furnish, or secure with glass. Double-glazed windows have two panes of glass for a higher R-value.

hydroelectricity (hydropower) Energy harnessed from flowing water. In some hydroelectric facilities, water is collected behind a dam and then used to turn a turbine, generating electricity.

incandescent bulb Light and heat created by electricity heating up a metal filament in the light bulb until it is white hot.

insulate Prevent passage of hot or cold air, electricity, or sound into or out of an area by surrounding it with non conducting material (insulation).

kilowatt A unit of power equal to one thousand watts. A kilowatt would run ten 100-watt light bulbs.

kilowatt-hour (kWh) The amount of energy expended by using one kilowatt for one hour. The US consumes 2.6 trillion kilowatt-hours a year.

landfill Method of solid waste disposal in which refuse is buried between layers of dirt. Layers fill in or reclaim low-lying ground.

lumen The unit of light emitted by one candle.

natural gas Underground deposits of gases consisting of 50% to 90% methane (CH_4) and small amounts of heavier gaseous hydrocarbon compounds such as propane (C_3H_8) and butane (C_4H_{10}).

photovoltaic An electric current produced by light striking a metal; the direct conversion of radiant energy into electrical energy.

pollutant A harmful substance deposited in the air, water, or land that leads to dirty, impure, or unhealthy conditions. Pollution of air, water, and land threatens human health, wildlife, and plants.

R-value The measure of a material's ability to resist the flow of heat. The higher the R-value, the better the insulation.

smog A term originally coined from the words smoke and fog; now applied also to the photochemical haze produced by the action of sun and atmosphere on automobile and industrial exhausts.

solar energy Direct radiant energy from the sun plus indirect forms of energy produced when solar energy interacts with the earth.

therm A unit of heat equal to 100,000 Btus.

turbine A bladed machine containing magnets that turns when air, water, or steam passes through it, generating electricity.

watt: A unit of power, or rate at which electrical work is done.

weather-stripping A narrow piece of material made of plastic, rubber, felt, or metal installed around doors and windows to seal.

ISSUES AND INFORMATION